The Mastery of the Air

William J. Claxton

Table of Contents

Table of Contents

The Mastery of the Air

William J. Claxton

PREFACE

This book makes no pretence of going minutely into the technical and scientific sides of human flight: rather does it deal mainly with the real achievements of pioneers who have helped to make aviation what it is to-day.

My chief object has been to arouse among my readers an intelligent interest in the art of flight, and, profiting by friendly criticism of several of my former works, I imagine that this is best obtained by setting forth the romance of triumph in the realms of an element which has defied man for untold centuries, rather than to give a mass of scientific principles which appeal to no one but the expert.

So rapid is the present development of aviation that it is difficult to keep abreast with the times. What is new to-day becomes old to-morrow. The Great War has given a tremendous impetus to the strife between the warring nations for the mastery of the air, and one can but give a rough and general impression of the achievements of naval and military airmen on the various fronts.

Finally, I have tried to bring home the fact that the fascinating progress of aviation should not be confined entirely to the airman and constructor of air-craft; in short, this progress is not a retord of events in which the mass of the nation have little personal concern, but of a movement in which each one of us may take an active and intelligent part.

I have to thank various aviation firms, airmen, and others who have kindly come to my assistance, either with the help of valuable information or by the loan of photographs. In particular, my thanks are due to the Royal Flying Corps and Royal Naval Air Service for permission to reproduce illustrations from their two publications on the work and training of their respective corps; to the Aeronautical Society of Great Britain; to Messrs. C. G. Spencer Sons, Highbury; The Sopwith Aviation Company, Ltd.; Messrs. A. V. Roe Co., Ltd.; The Gnome Engine Company; The Green Engine Company; Mr. A. G. Gross (Geographia, Ltd.); and M. Bleriot; for an exposition of the internal-combustion engine I have drawn on Mr. Horne's The Age of Machinery.

PART I–BALLOONS AND AIR–SHIPS

CHAPTER I. Man's Duel with Nature

Of all man's great achievements none is, perhaps, more full of human interest than are those concerned with flight. We regard ourselves as remarkable beings, and our wonderful discoveries in science and invention induce us to believe we are far and away the cleverest of all the living creatures in the great scheme of Creation. And yet in the matter of flight the birds beat us; what has taken us years of education, and vast efforts of intelligence, foresight, and daring to accomplish, is known by the tiny fledglings almost as soon as they come into the world.

It is easy to see why the story of aviation is of such romantic interest. Man has been exercising his ingenuity, and deliberately pursuing a certain train of thought, in an attempt to harness the forces of Nature and compel them to act in what seems to be the exact converse of Nature's own arrangements.

One of the mysteries of Nature is known as the FORCE OF GRAVITY. It is not our purpose in this book to go deeply into a study of gravitation; we may content ourselves with the statement, first proved by Sir Isaac Newton, that there is an invisible force which the Earth exerts on all bodies, by which it attracts or draws them towards itself. This property does not belong to the Earth alone, but to all matter—all matter attracts all other matter. In discussing the problems of aviation we are concerned mainly with the mutual attraction of The Earth and the bodies on or near its surface; this is usually called TERRESTRIAL gravity.

It has been found that every body attracts very other body with a force directly proportionate to its mass. Thus we see that, if every particle in a mass exerts its attractive influence, the more particles a body contains the greater will be the attraction. If a mass of iron be dropped to the ground from the roof of a building at the same time as a cork of similar size, the iron and the cork would, but for the retarding effect of the air, fall to the ground together, but the iron would strike the ground with much greater force than the cork. Briefly stated, a body which contains twice as much matter as another is attracted or drawn towards the centre of the Earth with twice the force of that other; if the mass be five times as great, then it will be attracted with five times the force, and so on.

It is thus evident that the Earth must exert an overwhelming attractive force on all bodies on or near its surface. Now, when man rises from the ground in an aeroplane he is counter–acting this force by other forces.

A short time ago the writer saw a picture which illustrated in a very striking manner man's struggle with Nature. Nature was represented as a giant of immense stature and strength, standing on a globe with outstretched arms, and in his hands were shackles of great size. Rising gracefully from the earth, immediately in front of the giant, was an airman seated in a modern flying–machine, and on his face was a happy–go–lucky look as though he were delighting in the duel between him and the giant. The artist had drawn the picture so skilfully that one could imagine the huge, knotted fingers grasping the shackles were itching to bring the airman within their clutch. The picture was entitled "MAN TRIUMPHANT"

No doubt many of those who saw that picture were reminded of the great sacrifices made by man in the past. In the wake of the aviator there are many memorial stones of mournful significance.

It says much for the pluck and perseverance of aviators that they have been willing to run the great risks which ever accompany their efforts. Four years of the Great War have shown how splendidly airmen have risen to the great demands made upon them. In dispatch after dispatch from the front, tribute has been paid to the gallant and devoted work of the Royal Flying Corps and the Royal Naval Air Service. In a long and bitter struggle British airmen have gradually asserted their supremacy in the air. In all parts of the globe, in Egypt, in Mesopotamia, in Palestine, in Africa, the airman has been an indispensable adjunct of the fighting forces. Truly it may be said that mastery of the air is the indispensable factor of final victory.

CHAPTER II. The French Paper–maker who Invented the Balloon

In the year 1782 two young Frenchmen might have been seen one winter night sitting over their cottage fire, performing the curious experiment of filling paper bags with smoke, and letting them rise up towards the ceiling. These young men were brothers, named Stephen and Joseph Montgolfier, and their experiments resulted in the invention of the balloon.

The brothers, like all inventors, seem to have had enquiring minds. They were for ever asking the why and the wherefore of things. "Why does smoke rise?" they asked. "Is there not some strange power in the atmosphere which makes the smoke from chimneys and elsewhere rise in opposition to the force of gravity? If so, cannot we discover this power, and apply it to the service of mankind?"

We may imagine that such questions were in the minds of those two French paper–makers, just as similar questions were in the mind of James Watt when he was discovering the power of steam. But one of the most important attributes of an inventor is an infinite capacity for taking pains, together with great patience.

And so we find the two brothers employing their leisure in what to us would, be a childish pastime, the making of paper balloons. The story tells us that their room was filled with smoke, which issued from the windows as though the house were on fire. A neighbour, thinking such was the case, rushed in, but, on being assured that nothing serious was wrong, stayed to watch the tiny balloons rise a little way from the thin tray which contained the fire that made the smoke with which the bags were filled. The experiments were not altogether successful, however, for the bags rarely rose more than a foot or so from the tray. The neighbour suggested that they should fasten the thin tray on to the bottom of the bag, for it was thought that the bags would not ascend higher because the smoke became cool; and if the smoke were imprisoned within the bag much better results would be obtained. This was done, and, to the great joy of the brothers and their visitor, the bag at once rose quickly to the ceiling.

But though they could make the bags rise their great trouble was that they did not know the cause of this ascent. They thought, however, that they were on the eve of some great discovery, and, as events proved, they were not far wrong. For a time they imagined that the fire they had used generated some special gas, and if they could find out the nature of this gas, and the means of making it in large quantities, they would be able to add to their success.

Of course, in the light of modern knowledge, it seems strange that the brothers did not know that the reason the bags rose, was not because of any special gas being used, but owing to the expansion of air under the influence of heat, whereby hot air tends to rise.

Every schoolboy above the age of twelve knows that hot air rises upwards in the atmosphere, and that it continues to rise until its temperature has become the same as that of the surrounding air.

The next experiment was to try their bags in the open air. Choosing a calm, fine day, they made a fire similar to that used in their first experiments, and succeeded in making the bag rise nearly 100 feet. Later on, a much larger craft was built, which was equally successful.

And now we must leave the experiments of the Montgolfiers for a moment, and turn to the discovery of hydrogen gas by Henry Cavendish, a well–known London chemist. In 1766 Cavendish proved conclusively that hydrogen gas was not more than one–seventh the weight of ordinary air. It at once occurred to Dr. Black, of Glasgow, that if a thin bag could be filled with this light gas it would rise in the air; but for various reasons his experiments did not yield results of a practical nature for several years.

Some time afterwards, about a year before the Montgolfiers commenced their experiments which we have already described, Tiberius Cavallo, an Italian chemist, succeeded in making, with hydrogen gas, soap–bubbles which rose in the air. Previous to this he had experimented with bladders and paper bags; but the bladders he found too heavy, and the paper too porous.

It must not be thought that the Montgolfiers experimented solely with hot air in the inflation of their balloons. At one time they used steam, and, later on, the newly–discovered hydrogen gas; but with both these agents they were unsuccessful. It can easily be seen why steam was of no use, when we consider that paper was employed; hydrogen, too, owed its lack of success to the same cause for the porosity of the paper allowed the gas to escape quickly.

It is said that the name "balloon" was given to these paper craft because they resembled in shape a large spherical vessel used in chemistry, which was known by that name. To the brothers Montgolfier belongs the honour of having given the name to this type of aircraft, which, in the two succeeding centuries, became so popular.

After numerous experiments the public were invited to witness the inflation of a particularly huge balloon, over 30 feet in diameter. This was accomplished over a fire made of wool and straw. The ascent was successful, and the balloon, after rising to a height of some 7000 feet, fell to earth about two miles away.

It may be imagined that this experiment aroused enormous interest in Paris, whence the news rapidly spread over all France and to Britain. A Parisian scientific society invited Stephen Montgolfier to Paris in order that the citizens of the metropolis should have their imaginations excited by seeing the hero of these remarkable experiments. Montgolfier was not a rich man, and to enable him to continue his experiments the society granted him a considerable sum of money. He was then enabled to construct a very fine balloon, elaborately decorated and painted, which ascended at Versailles in the presence of the Court.

To add to the value of this experiment three animals were sent up in a basket attached to the balloon. These were a sheep, a cock, and a duck. All sorts of guesses were made as to what would be the fate of the "poor creatures". Some people imagined that there was little or no air in those higher regions and that the animals would choke; others said they would be frozen to death. But when the balloon descended the cock was seen to be strutting about in his usual dignified way, the sheep was chewing the cud, and the duck was quacking for water and worms.

At this point we will leave the work of the brothers Montgolfier. They had succeeded in firing the imagination of nearly every Frenchman, from King Louis down to his humblest subject. Strange, was it not, though scores of millions of people had seen smoke rise, and clouds float, for untold centuries, yet no one, until the close of the eighteenth century, thought of making a balloon?

The learned Franciscan friar, Roger Bacon, who lived in the thirteenth century, seems to have thought of the possibility of producing a contrivance that would float in air. His idea was that the earth's atmosphere was a "true fluid", and that it had an upper surface as the ocean has. He quite believed that on this upper surface—subject, in his belief, to waves similar to those of the sea—an air–ship might float if it once succeeded in rising to the required height. But the difficulty was to reach the surface of this aerial sea. To do this he proposed to make a large hollow globe of metal, wrought as thin as the skill of man could make it, so that it might be as light as possible, and this vast globe was to be filled with "liquid fire". Just what "liquid fire" was, one cannot attempt to explain, and it is doubtful if Bacon himself had any clear idea. But he doubtless thought of some gaseous substance lighter than air, and so he would seem to have, at least, hit upon the principle underlying the construction of the modern balloon. Roger Bacon had ideas far in advance of his time, and his experiments made such an impression of wonder on the popular mind that they were believed to be wrought by black magic, and the worthy monk was classed among those who were supposed to be in league with Satan.

CHAPTER III. The First Man to Ascend in a Balloon

The safe descent of the three animals, which has already been related, showed the way for man to venture up in a balloon. In our time we marvel at the daring of modern airmen, who ascend to giddy heights, and, as it were, engage in mortal combat with the demons of the air. But, courageous though these deeds are, they are not more so than those of the pioneers of ballooning.

In the eighteenth century nothing was known definitely of the conditions of the upper regions of the air, where, indeed, no human being had ever been; and though the frail Montgolfier balloons had ascended and descended with no outward happenings, yet none could tell what might be the risk to life in committing oneself to an ascent. There was, too, very special danger in making an ascent in a hot–air balloon. Underneath the huge envelope was suspended a brazier, so that the fabric of the balloon was in great danger of catching fire.

It was at first suggested that two French criminals under sentence of death should be sent up, and, if they made a safe descent, then the way would be open for other aeronauts to venture aloft. But everyone interested in aeronautics in those days saw that the man who first traversed the unexplored regions of the air would be held in high honour, and it seemed hardly right that this honour should fall to criminals. At any rate this was the view of M. Pilatre de Rozier, a French gentleman, and he determined himself to make the pioneer ascent.

De Rozier had no false notion of the risks he was prepared to run, and he superintended with the greatest care the construction of his balloon. It was of enormous size, with a cage slung underneath the brazier for heating the air. Befors making his free ascent De Rozier made a trial ascent with the balloon held captive by a long rope.

At length, in November, 1783, accompanied by the Marquis d'Arlandes as a passenger, he determined to venture. The experiment aroused immense excitement all over France, and a large concourse of people were gathered together on the outskirts of Paris to witness the risky feat. The balloon made a perfect ascent, and quickly reached a height of about half a mile above sea–level. A strong current of air in the upper regions caused the balloon to take an opposite direction from that intended, and the aeronauts drifted right over Paris. It would have gone hard with them if they had been forced to descend in the city, but the craft was driven by the wind to some distance beyond the suburbs and they alighted quite safely about six miles from their starting–point, after having been up in the air for about half an hour.

Their voyage, however, had by no means been free from anxiety. We are told that the fabric of the balloon repeatedly caught fire, which it took the aeronauts all their time to extinguish. At times, too, they came down perilously near to the Seine, or to the housetops of Paris, but after the most exciting half–hour of their lives they found themselves once more on Mother Earth.

Here we must make a slight digression and speak of the invention of the hydrogen, or gas, balloon. In a previous chapter we read of the discovery of hydrogen gas by Henry Cavendish, and the subsequent experiments with this gas by Dr. Black, of Glasgow. It was soon decided to try to inflate a balloon with this "inflammable air"—as the newly–discovered gas was called—and with this end in view a large public subscription was raised in France to meet the heavy expenses entailed in the venture. The work was entrusted to a French scientist, Professor Charles, and two brothers named Robert.

It was quickly seen that paper, such as was used by the Montgolfiers, was of little use in the construction of a gas balloon, for the gas escaped. Accordingly the fabric was made of silk and varnished with a solution of india–rubber and turpentine. The first hydrogen balloon was only about 13 feet in diameter, for in those early days the method of preparing hydrogen was very laborious and costly, and the constructors thought it advisable not to spend too much money over the initial experiments, in case they should be a failure.

In August, 1783—an eventful year in the history of aeronautics— the first gas–inflated balloon was sent up, of course unaccompanied by a passenger. It shot up high in the air much more rapidly than Montgolfier's hot–air balloon had done, and was soon beyond the clouds. After a voyage of nearly an hour's duration it descended in a field some 15 miles away. We are told that some peasants at work near by fled in the greatest alarm at this strange monster which settled in their midst. An old print shows them cautiously approaching the balloon as it lay heaving on the ground, stabbing it with pitchforks, and beating it with flails and sticks. The story goes that one of the alarmed farmers poured a charge of shot into it with his gun, no doubt thinking that he had effectually silenced the panting demon contained therein. To prevent such unseemly occurrences in the future the French Government found it necessary to warn the people by proclamation that balloons were perfectly harmless objects, and that the experiments would be repeated.

We now have two aerial craft competing for popular favour: the Montgolfier hot–air balloon and the "Charlier" or gas–inflated balloon. About four months after the first trial trip of the latter the inventors decided to ascend in a specially–constructed hydrogen–inflated craft. This balloon, which was 27 feet in diameter, contained nearly all the features of the modern balloon. Thus there was a valve at the top by means of which the gas could be let out as desired; a cord net covered the whole fabric, and from the loop which it formed below the neck of the balloon a car was suspended; and in the car there was a quantity of ballast which could be cast overboard when necessary.

It may be imagined that this new method of aerial navigation had thoroughly aroused the excitability of the French nation, so that thousands of people were met together just outside Paris on the 17th December to see Professor Charles and his mechanic, Robelt, ascend in their new craft. The ascent was successful in every way; the intrepid aeronauts, who carried a barometer, found that they had quickly reached an altitude of over a mile.

After remaining aloft for nearly two hours they came down. Professor Charles decided to ascend again, this time by himself, and with a much lighter load the balloon rose about two miles above sea–level. The temperature at this height became very low, and M. Charles was affected by violent pain in his right ear and jaw. During the voyage he witnessed the strange phenomenon of a double sunset; for, before the ascent, the sun had set behind the hills overshadowing the valleys, and when he rose above the hill–tops he saw the sun again, and presently saw it set again. There is no doubt that the balloon would have risen several thousand feet higher, but the professor thought it would burst, and he opened the valve, eventually making a safe descent about 7 miles from his starting–place.

England lagged behind her French neighbour's in balloon aeronautics—much as she has recently done in aviation—for a considerable time, and,it was not till August of the following year (1784) that the first balloon ascent was made in Great Britain, by Mr. J. M. Tytler. This took place at Edinburgh in a fire balloon. Previous to this an Italian, named Lunardi, had in November, 1783, dispatched from the Artillery Ground, in London, a small balloon made of oil–silk, 10 feet in diameter and weighing 11 pounds. This small craft was sent aloft at one o'clock, and came down, about two and a half hours later, in Sussex, about 48 miles from its starting–place.

In 1784 the largest balloon on record was sent up from Lyons. This immense craft was more than 100 feet in diameter, and stood about 130 feet high. It was inflated with hot air over a straw fire, and seven passengers were carried, including Joseph Montgolfier and Pilatre de Rozier.

But to return to de Rozier, whom we left earlier in the chapter, after his memorable ascent near Paris. This daring Frenchman decided to cross the Channel, and to prevent the gas cooling, and the balloon falling into the sea, he hit on the idea of suspending a small fire balloon under the neck of another balloon inflated with hydrogen gas. In the light of our modern knowledge of the highly–inflammable nature of hydrogen, we wonder how anyone could have attempted such an adventure; but there had been little experience of this newly–discovered gas in those days. We are not surprised to read that, when high in the air, there was an awful explosion and the brave aeronaut fell to the earth and was dashed to death.

CHAPTER IV. The First Balloon Ascent in England

It has been said that the honour of making the first ascent in a balloon from British soil must be awarded to Mr. Tytler. This took place in Scotland. In this chapter we will relate the almost romantic story of the first ascent made in England.

This was carried out successfully by Lunardi, the Italian of whom we have previously spoken. This young foreigner, who was engaged as a private secretary in London, had his interest keenly aroused by the accounts of the experiments being carried out in balloons in France, and he decided to attempt similar experiments in this country.

But great difficulties stood in his way. Like many other inventors and would–be airmen, he suffered from lack of funds to build his craft, and though people whom he approached for financial aid were sympathetic, many of them were unwilling to subscribe to his venture. At length, however, by indomitable perseverance, he collected enough money to defray the cost of building his balloon, and it was arranged that he should ascend from the Artillery Ground, London, in September, 1784.

His craft was a "Charlier"—that is, it was modelled after the hydrogen–inflated balloon built by Professor Charles—and it resembled in shape an enormous pear. A wide hoop encircled the neck of the envelope, and from this hoop the car was suspended by stout cordage.

It is said that on the day announced for the ascent a crowd of nearly 200,000 had assembled, and that the Prince of Wales was an interested spectator. Farmers and labourers and, indeed, all classes of people from the prince down to he humblest subject, were represented, and seldom had London's citizens been more deeply excited.

Many of them, however, were incredulous, especially when an insufficiency of gas caused a long delay before the balloon could be liberated. Fate seemed to be thwarting the plucky Italian at every step. Even at the last minute, when all arrangements had been perfected as far as was humanly possible, and the crowd was agog with excitement, it appeared probable that he would have to postpone the ascent.

It was originally intended that Lunardi should be accompanied by a passenger; but as there was a shortage of gas the balloon's lifting power was considerably lessened, and he had to take the trip with a dog and cat for companions. A perfect ascent was made, and in a few moments the huge balloon was sailing gracefully in a northerly direction over innumerable housetops.

This trip was memorable in another way. It was probably the only aerial cruise where a Royal Council was put off in order to witness the flight. It is recorded that George the Third was in conference with the Cabinet, and when news arrived in the Council Chamber that Lunardi was aloft, the king remarked: Gentlemen, we may resume our deliberations at pleasure, but we may never see poor Lunardi again!"

The journey was uneventful; there was a moderate northerly breeze, and the aeronaut attained a considerable altitude, so that he and his animals were in danger of frost–bite. Indeed, one of the animals suffered so severely from the effects of the cold that Lunardi skilfully descended low enough to drop it safely to earth, and then, throwing out ballast, once more ascended. He eventually came to earth near a Hertfordshire village about 30 miles to the north of London.

CHAPTER V. The Father of British Aeronauts

No account of the early history of English aeronautics could possibly be complete unless it included a description of the Nassau balloon, which was inflated by coal–gas, from the suggestion of Mr. Charles Green, who was one of Britain's most famous aeronauts. Because of his institution of the modern method of using coal–gas in a balloon, Mr. Green is generally spoken of as the Father of British Aeronautics. During the close of the eighteenth and the opening years of the nineteenth century there had been numerous ascents in Charlier balloons, both in Britain and on the Continent. It had already been discovered that hydrogen gas was highly dangerous and also expensive, and Mr. Green proposed to try the experiment of inflating a balloon with ordinary coal–gas, which had now become fairly common in most large towns, and was much less costly than hydrogen.

Critics of the new scheme assured the promoters that coal–gas would be of little use for a balloon, averring that it had comparatively little lifting power, and aeronauts could never expect to rise to any great altitude in such a balloon. But Green firmly believed that his theory was practical, and he put it to the test. The initial experiments quite convinced him that he was right. Under his superintendence a fine balloon about 80 feet high, built of silk, was made in South London, and the car was constructed to hold from fifteen to twenty passengers. When the craft was completed it was proposed to send it to Paris for exhibition purposes, and the inventor, with two friends, Messrs. Holland and Mason, decided to take it over the Channel by air. It is said that provisions were taken in sufficient quantities to last a fortnight, and over a ton of ballast was shipped.

The journey commenced in November, 1836, late in the afternoon, as the aeronauts had planned to cross the sea by night. A fairly strong north–west wind quickly bore them to the coast, and in less than an hour they found themselves over the lights of Calais. On and on they went, now and then entirely lost to Earth through being enveloped in dense fog; hour after hour went by, until at length dawn revealed a densely–wooded tract of country with which they were entirely unfamiliar. They decided to land, and they were greatly surprised to find that they had reached Weilburg, in Nassau, Germany. The whole journey of 500 miles had been made in eighteen hours.

Probably no British aeronaut has made more daring and exciting ascents than Mr. Green—unless it be a member of the famous Spencer family, of whom we speak in another chapter. It is said that Mr. Green went aloft over a thousand times, and in later years he was accompanied by various passengers who were making ascents for scientific purposes. His skill was so great that though he had numerous hairbreadth escapes he seldom suffered much bodily harm. He lived to the ripe old age of eighty–five.

CHAPTER VI. The Parachute

No doubt many of those who read this book have seen an aeronaut descend from a balloon by the aid of a parachute. For many years this performance has been one of the most attractive items on the programmes of fetes, galas, and various other outdoor exhibitions.

The word "parachute" has been almost bodily taken from the French language. It is derived from the French parer to parry, and chute a fall. In appearance a parachute is very similar to an enormous umbrella.

M. Blanchard, one of the pioneers of ballooning, has the honour of first using a parachute, although not in person. The first "aeronaut" to descend by this apparatus was a dog. The astonished animal was placed in a basket attached to a parachute, taken up in a balloon, and after reaching a considerable altitude was released. Happily for the dog the parachute acted quite admirably, and the animal had a graceful and gentle descent.

Shortly afterwards a well–known French aeronaut, M. Garnerin, had an equally satisfactory descent, and soon the parachute was used by most of the prominent aeronauts of the day. Mr. Cocking, a well–known balloonist, held somewhat different views from those of other inventors as to the best form of construction of parachutes. His idea was that a parachute should be very large and rather heavy in order to be able to support a great weight. His first descent from a great height was also his last. In 1837, accompanied by Messrs. Spencer and Green, he went up with his parachute, attached to the Nassau balloon. At a height of about a mile the parachute was liberated, but it failed to act properly; the inventor was cast headlong to earth, and dashed to death.

From time to time it has been thought that the parachute might be used for life–saving on the modern dirigible air–ship, and even on the aeroplane, and experiments have been carried out with that end in view. A most thrilling descent from an air–ship by means of a parachute was that made by Major Maitland, Commander of the British Airship Squadron, which forms part of the Royal Flying Corps. The descent took place from the Delta air–ship, which ascended from Farnborough Common. In the car with Major Maitland were the pilot, Captain Waterlow, and a passenger. The parachute was suspended from the rigging of the Delta, and when a height of about 2000 feet had been reached it was dropped over to the side of the car. With the dirigible travelling at about 20 miles an hour the major climbed over the car and seated himself in the parachute. Then it became detached from the Delta and shot downwards for about 200 feet at a terrific rate. For a moment or two it was thought that the opening apparatus had failed to work; but gradually the "umbrella" opened, and the gallant major had a gentle descent for the rest of the distance.

This experiment was really made in order to prove the stability of an air–ship after a comparatively great weight was suddenly removed from it. Lord Edward Grosvenor, who is attached to the Royal Flying Corps, was one of the eyewitnesses of the descent. In speaking of it he said: "We all think highly of Major Maitland's performance, which has shown how the difficulty of lightening an air–ship after a long flight can be surmounted. During a voyage of several hours a dirigible naturally loses gas, and without some means of relieving her of weight she might have to descend in a hostile country. Major Maitland has proved the practicability of members of an air–ship's crew dropping to the ground if the necessity arises."

A descent in a parachute has also been made from an aeroplane by M. Pegoud, the daring French airman, of whom we speak later. A certain Frenchman, M. Bonnet, had constructed a parachute which was intended to be used by the pilot of an aeroplane if on any occasion he got into difficulties. It had been tried in many ways, but, unfortunately for the inventor, he could get no pilot to trust himself to it. Tempting offers were made to pilots of world–wide fame, but either the risk was thought to be too great, or it was believed that no practical good would come of the experiment. At last the inventor approached M. Pegoud, who undertook to make the descent. This was accomplished from a great height with perfect safety. It seems highly probable that in the near future the parachute will form part of the equipment of every aeroplane and air–ship.

CHAPTER VII. Some British Inventors of Air–ships

The first Englishman to invent an air–ship was Mr. Stanley Spencer, head of the well–known firm of Spencer Brothers, whose worksare at Highbury, North London.

This firm has long held an honourable place in aeronautics, both in the construction of air–craft and in aerial navigation. Spencer Brothers claim to be the premier balloon manufacturers in the world, and, at the time of writing, eighteen balloons and two dirigibles lie in the works ready for use. In these works there may also be seen the frame of the famous Santos–Dumont air–ship, referred to later in this book.

In general appearance the first Spencer air–ship was very similar to the airship flown by Santos–Dumont; that is, there was the cigar–shaped balloon, the small engine, and the screw propellor for driving the craft forward.

But there was one very important distinction between the two air–ships. By a most ingenious contrivance the envelope was made so that, in the event of a large and serious escape of gas, the balloon would assume the form of a giant umbrella, and fall to earth after the manner of a parachute.

All inventors profit, or should profit, by the experience of others, whether such experience be gained by success or failure. It was found that Santos–Dumont's air–ship lost a considerable amount of gas when driven through the air, and on several occasions the whole craft was in great danger of collapse. To keep the envelope inflated as tightly as possible Mr. Spencer, by a clever contrivance, made it possible to force air into the balloon to replace the escaped gas.

The first Spencer air–ship was built for experimental purposes. It was able to lift only one person of light weight, and was thus a great contrast to the modern dirigible which carries a crew of thirty or forty people. Mr. Spencer made several exhibition flights in his little craft at the Crystal Palace, and so successful were they that he determined to construct a much larger craft.

The second Spencer air–ship, first launched in 1903, was nearly 100 feet long. There was one very important distinction between this and other air–ships built at that time: the propeller was placed in front of the craft, instead of at the rear, as is the case in most air–ships. Thus the craft was pulled through the air much after the manner of an aeroplane.

In the autumn of 1903 great enthusiasm was aroused in London by the announcement that Mr. Spencer proposed to fly from the Crystal Palace round the dome of St. Paul's Cathedral and back to his starting–place. This was a much longer journey than that made by Santos–Dumont when he won the Deutsch prize.

Tens of thousands of London's citizens turned out to witness the novel sight of a giant air–ship hovering over the heart of their city, and it was at once seen what enormous possibilities there were in the employment of such craft in time of war. The writer remembers well moving among the dense crowds and hearing everywhere such remarks as these:

"What would happen if a few bombs were thrown over the side of the air–ship?" "Will there be air–fleets in future, manned by the soldiers or sailors?" Indeed the uppermost thought in people's minds was not so much the possibility of Mr. Spencer being able to complete his journey successfully—nearly everyone recognized that air–ship construction had now advanced so far that it was only a matter of time for an ideal craft to be built—but that the coming of the air–ship was an affair of grave international importance.

The great craft, glistening in the sunlight, sailed majestically from the south, but when it reached the Cathedral it refused to turn round and face the wind. Try how he might, Mr. Spencer could not make any progress. It was a thrilling sight to witness this battle with the elements, right over the heart of the largest city in the world. At times the air–ship seemed to be standing quite still, head to wind. Unfortunately, half a gale had sprung up, and the 24–horse–power engine was quite incapable of conquering so stiff a breeze, and making its way home again. After several gallant attempts to circle round the dome, Mr. Spencer gave up in despair, and let the monster air–ship drift with the wind over the northern suburbs of the city until a favourable landing–place near Barnet was reached, where he descended.

The Spencer air–ships are of the non–rigid type. Spencer air–ship A comprises a gas vessel for hydrogen 88 feet long and 24 feet in diameter, with a capacity of 26,000 cubic feet. The framework is of polished ash wood, made in sections so that it can easily be taken to pieces and transported, and the length over all is 56 feet. Two propellers 7 feet 6 inches diameter, made of satin–wood, are employed to drive the craft, which is equipped with a Green engine of from 35 to 40 horse–power.

Spencer's air–ship B is a much larger vessel, being 150 feet long and 35 feet in diameter, with a capacity for hydrogen of 100,000 cubic feet. The framework is of steel and aluminium, made in sections, with cars for ten persons, including aeronauts, mechanics, and passengers. It is driven with two petrol aerial engines of from 50 to 60 horse–power.

About the time that Mr. Spencer was experimenting with his large air–ship, Dr. Barton, of Beckenham, was forming plans for an even larger craft. This he laid down in the spacious grounds of the Alexandra Park, to the north of London. An enormous shed was erected on the northern slopes of the park, but visitors to the Alexandra Palace, intent on a peep at the monster air–ship under construction, were sorely disappointed, as the utmost secrecy in the building of the craft was maintained.

The huge balloon was 43 feet in diameter and 176 feet long, with a gas capacity of 235,000 cubic feet. To maintain the external form of the envelope a smaller balloon, or compensator, was placed inside the larger one. The framework was of bamboo, and the car was attached by about eighty wire–cables. The wooden deck was about 123 feet in length. Two 50–horse–power engines drove four propellers, two of which were at either end.

The inventor employed a most ingenious contrivance to preserve the horizontal balance of the air–ship. Fitted, one at each end of the carriage, were two 50–gallon tanks. These tanks were connected with a long pipe, in the centre of which was a hand–pump. When the bow of the air–ship dipped, the man at the pump could transfer some of the water from the fore–tank to the after–tank, and the ship would right itself. The water could similarly be transferred from the after–tank to the fore–tank when the stern of the craft pointed downwards.

There were many reports, in the early months of 1905, that the air–ship was going to be brought out from the shed for its trial flights, and the writer, in common with many other residents in the vicinity of the park, made dozens of journeys to the shed in the expectation of seeing the mighty dirigible sail away. But for months we were doomed to disappointment; something always seemed to go wrong at the last minute, and the flight had to be postponed.

At last, in 1905, the first ascent took place. It was unsuccessful. The huge balloon, made of tussore silk, cruised about for some time, then drifted away with the breeze, and came to grief in landing.

A clever inventor of air–ships, a young Welshman, Mr. E. T. Willows, designed in 1910, an air–ship in which he flew from Cardiff to London in the dark—a distance of 139 miles. In the same craft he crossed the English Channel a little later.

Mr. Willows has a large shed in the London aerodrome at Hendon, and he is at present working there on a new air–ship. For some time he has been the only successful private builder of air–ships in Great Britain. The Navy possess a small Willows air–ship.

Messrs. Vickers, the famous builders of battleships, are giving attention to the construction of air–ships for the Navy, in their works at Walney Island, Barrow–in–Furness. This firm has erected an enormous shed, 540 feet long, 150 feet broad, and 98 feet high. In this shed two of the largest air–ships can be built side by side. Close at hand is an extensive factory for the production of hydrogen gas.

At each end of the roof are towers from which the difficult task of safely removing an air–ship from the shed can be directed.

At the time of writing, the redoubtable DORA (Defence of the Realm Act) forbids any but the vaguest references to what is going forward in the way of additions to our air forces. But it may be stated that air–ships are included in the great constructive programme now being carried out. It is not long since the citizens of Glasgow were treated to the spectacle of a full–sized British "Zep" circling round the city prior to her journey south, and so to regions unspecified. And use, too, is being found by the naval arm for that curious hybrid the "Blimp", which may be described as a cross between an aeroplane and an air–ship.

CHAPTER VIII. The First Attempts to Steer a Balloon

For nearly a century after the invention of the Montgolfier and Charlier balloons there was not much progress made in the science of aeronautics. True, inventors such as Charles Green suggested and carried out new methods of inflating balloons, and scientific observations of great importance were made by balloonists both in Britain and on the Continent. But in the all–important work of steering the huge craft, progress was for many years practically at a standstill. All that the balloonist could do in controlling his balloon was to make it ascend or descend at will; he could not guide its direction of flight. No doubt pioneers of aeronautics early turned their attention to the problem of providing some apparatus, or some method, of steering their craft. One inventor suggested the hoisting of a huge sail at the side of the envelope; but when this was done the balloon simply turned round with the sail to the front. It had no effect on the direction of flight of the balloon. "Would not a rudder be of use?" someone asked. This plan was also tried, but was equally unsuccessful.

Perhaps some of us may wonder how it is that a rudder is not as serviceable on a balloon as it is on the stern of a boat. Have you ever found yourself in a boat on a calm day, drifting idly down stream, and going just as fast as the stream goes? Work the rudder how you may, you will not alter the boat's course. But supposing your boat moves faster than the stream, or by some means or other is made to travel slower than the current, then your rudder will act, and you may take what direction you will.

It was soon seen that if some method could be adopted whereby the balloon moved through the air faster or slower than the wind, then the aeronaut would be able to steer it. Nowadays a balloon's pace can be accelerated by means of a powerful motor–engine, but the invention of the petrol–engine is very recent. Indeed, the cause of the long delay in the construction of a steerable balloon was that a suitable engine could not be found. A steam–engine, with a boiler of sufficient power to propel a balloon, is so heavy that it would require a balloon of impossible size to lift it.

One of the first serious attempts to steer a balloon by means of engine power was that made by M. Giffard in 1852. Giffard's balloon was about 100 feet long and 40 feet in diameter, and resembled in shape an elongated cigar. A 3–horse–power steam–engine, weighing nearly 500 pounds, was provided to work a propeller, but the enormous weight was so great in proportion to the lifting power of the balloon that for a time the aeronaut could not leave the ground. After several experiments the inventor succeeded in ascending, when he obtained a speed against the wind of about 6 miles an hour.

A balloon of great historical interest was that invented by Dtipuy du Lonie, in the year 1872. Instead of using steam he employed a number of men to propel the craft, and with this air–ship he hoped to communicate with the besieged city of Paris.

His greatest speed against a moderate breeze was only about 5 miles an hour, and the endurance of the men did not allow of even this speed being kept up for long at a time.

Dupuy foreshadowed the construction of the modern dirigible air–ship by inventing a system of suspension links which connected the car to the envelope; and he also used an internal ballonet similar to those described in Chapter X.

In the year 1883 Tissandier invented a steerable balloon which was fitted with an electric motor of 1 1/2 horse–power. This motor drove a propeller, and a speed of about 8 miles an hour was attained. It is interesting to contrast the power obtained from this engine with that of recent Zeppelin air–ships, each of which is fitted with three or four engines, capable of producing over 800 horse–power.

The first instance on record of an air–ship being steered back to its starting–point was that of La France. This air–craft was the invention of two French army captains, Reynard and Krebs. By special and much–improved electric motors a speed of about 14 miles an hour was attained.

Thus, step by step, progress was made; but notwithstanding the promising results it was quite evident that the engines were far too heavy in proportion to the power they supplied. At length, however, the internal–combustion engine, such as is used in motor–cars, arrived, and it became at last possible to solve the great problem of constructing a really–serviceable, steerable balloon.

CHAPTER IX. The Strange Career of Count Zeppelin

In Berlin, on March 8, 1917, there passed away a man whose name will be remembered as long as the English language is spoken. For Count Zeppelin belongs to that little band of men who giving birth to a work of genius have also given their names to the christening of it; and so the patronymic will pass down the ages.

In the most sinister sense of the expression Count Zeppelin may be said to have left his mark deep down upon the British race. In course of time many old scores are forgiven and forgotten, but the Zeppelin raids on England will survive, if only as a curious failure. Their failure was both material and moral. Anti–aircraft guns and our intrepid airmen brought one after another of these destructive monsters blazing to the ground, and their work of "frightfulness" was taken up by the aeroplane; while more lamentable still was the failure of the Zeppelin as an instrument of terror to the civil population. In the long list of German miscalculations must be included that which pictured the victims of bombardment from the air crying out in terror for peace at any price.

Before the war Count Zeppelin was regarded by the British public as rather a picturesque personality. He appeared in the romantic guise of the inventor struggling against difficulties and disasters which would soon have overwhelmed a man of less resolute character. Even old age was included in his handicap, for he was verging on seventy when still arming against a sea of troubles.

The ebb and flow of his fortunes were followed with intense interest in this country, and it is not too much to say that the many disasters which overtook his air–ships in their experimental stages were regarded as world–wide calamities.

When, finally, the Count stood on the brink of ruin and the Kaiser stepped forward as his saviour, something like a cheer went up from the British public at this theatrical episode. Little did the audience realize what was to be the outcome of the association between these callous and masterful minds.

And now for a brief sketch of Count Zeppelin's life–story. He was born in 1838, in a monastery on an island in Lake Constance. His love of adventure took him to America, and when he was about twenty–five years of age he took part in the American Civil War. Here he made his first aerial ascent in a balloon belonging to the Federal army, and in this way made that acquaintance with aeronautics which became the ruling passion of his life.

After the war was over he returned to Germany, only to find another war awaiting him—the Austro–Prussian campaign. Later on he took part in the Franco–Prussian War, and in both campaigns he emerged unscathed.

But his heart was not in the profession of soldiering. He had the restless mind of the inventor, and when he retired, a general, after twenty years' military service, he was free to give his whole attention to his dreams of aerial navigation. His greatest ambition was to make his country pre–eminent in aerial greatness.

Friends to whom he revealed his inmost thoughts laughed at him behind his back, and considered that he was "a little bit wrong in his head". Certainly his ideas of a huge aerial fleet appeared most extravagant, for it must be remembered that the motor–engine had not then arrived, and there appeared no reasonable prospect of its invention.

Perseverance, however, was the dominant feature of Count Zeppelin's character; he refused to be beaten. His difficulties were formidable. In the first place, he had to master the whole science of aeronautics, which implies some knowledge of mechanics, meteorology, and electricity. This in itself was no small task for a man of over fifty years of age, for it was not until Count Zeppelin had retired from the army that he began to study these subjects at all deeply.

The next step was to construct a large shed for the housing of his air–ship, and also for the purpose of carrying out numerous costly experiments. The Count selected Friedrichshafen, on the shores of Lake Constance, as his head–quarters. He decided to conduct his experiments over the calm waters of the lake, in order to lessen the effects of a fall. The original shed was constructed on pontoons, and it could be turned round as desired, so that the air–ship could be brought out in the lee of any wind from whatsoever quarter it came.

It is said that the Count's private fortune of about L25,000 was soon expended in the cost of these works and the necessary experiments. To continue his work he had to appeal for funds to all his friends, and also to all patriotic Germans, from the Kaiser downwards.

At length, in 1908, there came a turning–point in his fortunes. The German Government, which had watched the Count's progress with great interest, offered to buy his invention outright if he succeeded in remaining aloft in one of his dirigibles for twenty–four hours.

The Count did not quite succeed in his task, but he aroused the great interest of the whole German nation, and a Zeppelin fund was established, under the patronage of the Kaiser, in every town and city in the Fatherland. In about a month the fund amounted to over L300,000. With this sum the veteran inventor was able to extend his works, and produce air–ship after air–ship with remarkable rapidity.

When, war broke out it is probable that Germany possessed at least thirteen air–ships which had fulfilled very difficult tests. One had flown 1800 miles in a single journey. Thus the East Coast of England, representing a return journey of less than 600 miles was well within their range of action.

CHAPTER X. A Zeppelin Air–ship and its Construction

After the Zeppelin fund had brought in a sum of money which probably exceeded all expectations, a company was formed for the construction of dirigibles in the Zeppelin works on Lake Constance, and in 1909 an enormous air–ship was produced.

In shape a Zeppelin dirigible resembled a gigantic cigar, pointed at both ends. If placed with one end on the ground in Trafalgar Square, London, its other end would be nearly three times the height of the Nelson Column, which, as you may know, is 166 feet.

From the diagram here given, which shows a sectional view of a typical Zeppelin air–ship, we may obtain a clear idea of the main features of the craft. From time to time, during the last dozen years or so, the inventor has added certain details, but the main features as shown in the illustration are common to all air–craft of this type.

Zeppelin L1 was 525 feet in length, with a diameter of 50 feet. Some idea of the size may be obtained through the knowledge that she was longer than a modern Dreadnought. The framework was made of specially light metal, aluminium alloy, and wood. This framework, which was stayed with steel wire, maintained the shape and rigidity of her gas–bags; hence vessels of this type are known as RIGID air–ships. Externally the hull was covered with a waterproof fabric.

Though, from outside, a rigid air–ship looks to be all in one piece, within it is divided into numerous compartments. In Zeppelin L1 there were eighteen separate compartments, each of which contained a balloon filled with hydrogen gas. The object of providing the vessel with these small balloons, or ballonets, all separate from one another, was to prevent the gas collecting all at one end of the ship as the vessel travelled through the air. Outside the ballonets there was a ring–shaped, double bottom, containing non–inflammable gas, and the whole was enclosed in rubber–coated fabric.

The crew and motors were carried in cars slung fore and aft. The ship was propelled by three engines, each of 170 horse–power. One engine was placed in the forward car, and the two others in the after car. To steer her to right or left, she had six vertical planes somewhat resembling box–kites, while eight horizontal planes enabled her to ascend or descend.

In Zeppelin L2, which was a later type of craft, there were four motors capable of developing 820 horse–power. These drove four propellers, which gave the craft a speed of about 45 miles an hour.

The cars were connected by a gangway built within the framework. On the top of the gas–chambers was a platform of aluminium alloy, carrying a 1–pounder gun, and used also as an observation station. It is thought that L1 was also provided with four machine–guns in her cars.

Later types of Zeppelins were fitted with a "wireless" installation of sufficient range to transmit and receive messages up to 350 miles. L1 could rise to the height of a mile in favourable weather, and carry about 7 tons over and above her own weight.

Even when on ground the unwieldy craft cause many anxious moments to the officers and mechanics who handle them. Two of the line have broken loose from their anchorage in a storm and have been totally destroyed. Great difficulty is also experienced in getting them in and out of their sheds. Here, indeed, is a contrast with the ease and rapidity with which an aeroplane is removed from its hangar.

It was maintained by the inventor that, as the vessel is rigid, and therefore no pressure is required in the gas–chamber to maintain its shape, it will not be readily vulnerable to projectiles. But the Count did not foresee that the very "frightfulness" of his engine of war would engender counter–destructives. In a later chapter an account will be given of the manner in which Zeppelin attacks upon these islands were gradually beaten off by the combined efforts of anti–aircraft guns and aeroplanes. To the latter, and the intrepid pilots and fighters, is due the chief credit for the final overthrow of the Zeppelin as a weapon of offence. Both the British and French airmen in various brilliant sallies succeeded in gradually breaking up and destroying this Armada of the Air; and the Zeppelin was forced back to the one line of work in which it has proved a success, viz., scouting for the German fleet in the few timid sallies it has made from home ports.

CHAPTER XI. The Semi–rigid Air–ship

Modern air–ships are of three general types: RIGID, SEMI–RIGID, and NON–RIGID. These differ from one another, as the names suggest, in the important feature, the RIGIDITY, NON–RIGIDITY, and PARTIAL RIGIDITY of the gas envelope.

Hitherto we have discussed the RIGID type of vessel with which the name of Count Zeppelin is so closely associated. This vessel is, as we have seen, not dependent for its form on the gas–bag, but is maintained in permanent shape by means of an aluminium framework. A serious disadvantage to this type of craft is that it lacks the portability necessary for military purposes. It is true that the vessel can be taken to pieces, but not quickly. The NON–RIGID type, on the other hand, can be quickly deflated, and the parts of the car and engine can be readily transported to the nearest balloon station when occasion requires.

In the SEMI–RIGID type of air–ship the vessel is dependent for its form partly on its framework and partly on the form of the gas envelope. The under side of the balloon consists of a flat rigid framework, to which the planes are attached, and from which the car, the engine, and propeller are suspended.

As the rigid type of dirigible is chiefly advocated in Germany, so the semi–rigid craft is most popular in France. The famous Lebaudy air–ships are good types of semi–rigid vessels. These were designed for the firm of Lebaudy Freres by the well–known French engineer M. Henri Julliot.

In November, 1902, M. Julliot and M. Surcouf completed an air–ship for M. Lebaudy which attained a speed of nearly 25 miles an hour. The craft, which was named Lebaudy I, made many successful voyages, and in 1905 M. Lebaudy offered a second vessel,

Lebaudy II, to the French Minister of War, who accepted it for the French nation, and afterwards decided to order another dirigible, La Patrie, of the same type. Disaster, however, followed these air–ships. Lebaudy I was torn from its anchorage during a heavy gale in 1906, and was completely wrecked. La Patrie, after travelling in 1907 from Paris to Verdun, in seven hours, was, a few days later, caught in a gale, and the pilot was forced to descend. The wind, however, was so strong that 200 soldiers were unable to hold down the unwieldy craft, and it was torn from their hands. It sailed away in a north–westerly direction over the Channel into England, and ultimately disappeared into the North Sea, where it was subsequently discovered some days after the accident.

Notwithstanding these disasters the French military authorities ordered another craft of the same type, which was afterwards named the Republique. This vessel made a magnificent flight of six and a half hours in 1908, and it was considered to have quite exceptional features, which eclipsed the previous efforts of Messrs. Julliot and Lebaudy.

Unfortunately, however, this vessel was wrecked in a very terrible manner. While out cruising with a crew of four officers one of the propeller blades was suddenly fractured, and, flying off with immense force, it entered the balloon, which it ripped to pieces. The majestic craft crumpled up and crashed to the ground, killing its crew in its fall.

In the illustration facing p. 17, of a Lebaudy air–ship, we have a good type of the semi–rigid craft. In shape it somewhat resembles an enormous porpoise, with a sharply–pointed nose. The whole vessel is not as symmetrical as a Zeppelin dirigible, but its inventors claim that the sharp prow facilitates the steady displace ment of the air during flight. The stern is rounded so as to provide sufficient support for the rear planes.

Two propellers are employed, and are fixed outside the car, one on each side, and almost in the centre of the vessel. This is a some what unusual arrangement. Some inventors, such as Mr. Spencer, place the propellers at the prow, so that the air–ship is DRAWN along; others prefer the propeller at the stern, whereby the craft is PUSHED along; but M. Julliot chose the central position, because there the disturbance of the air is smallest.

The body of the balloon is not quite round, for the lower part is flattened and rests on a rigid frame from which the car is suspended. The balloon is divided into three compartments, so that the heavier air does not move to one part of the balloon when it is tilted.

In the picture there is shown the petrol storage–tank, which is suspended immediately under the rear horizontal plane, where it is out of danger of ignition from the hot engine placed in the car.

CHAPTER XII. A Non–rigid Balloon

Hitherto we have described the rigid and semi–rigid types of air–ships. We have seen that the former maintains its shape without assistance from the gas which inflates its envelope and supplies the lifting power, while the latter, as its name implies, is dependent for its form partly on the flat rigid framework to which the car is attached, and partly on the gas balloon.

We have now to turn our attention to that type of craft known as a NON–RIGID BALLOON. This vessel relies for its form ENTIRELY upon the pressure of the gas, which keeps the envelope distended with sufficient tautness to enable it to be driven through the air at a considerable speed.

It will at once be seen that the safety of a vessel of this type depends on the maintenance of the gas pressure, and that it is liable to be quickly put out of action if the envelope becomes torn. Such an occurrence is quite possible in war. A well–directed shell which pierced the balloon would undoubtedly be disastrous to air–ship and crew. For this reason the non–rigid balloon does not appear to have much future value as a fighting ship. But, as great speed can be obtained from it, it seems especially suited for short overland voyages, either for sporting or commercial purposes. One of its greatest advantages is that it can be easily deflated, and can be packed away into a very small compass.

A good type of the non–rigid air–ship is that built by Major Von Parseval, which is named after its inventor. The Parseval has been described as "a marvel of modern aeronautical construction", and also as "one of the most perfect expressions of modern aeronautics, not only on account of its design, but owing to its striking efficiency.

The balloon has the elongated form, rounded or pointed at one end, or both ends, which is common to most air–ships. The envelope is composed of a rubber–texture fabric, and externally it is painted yellow, so that the chemical properties of the sun's rays may not injure the rubber. There are two smaller interior balloons, or COMPENSATORS, into which can be pumped air by means of a mechanically–driven fan or ventilator, to make up for contraction of the gas when descending or meeting a cooler atmosphere. The compensators occupy about one–quarter of the whole volume.

To secure the necessary inclination of the balloon while in flight, air can be transferred from one of the compensators, say at the fore end of the ship, into the ballonet in the aft part. Suppose it is desired to incline the bow of the craft upward, then the ventilating fan would DEFLATE the fore ballonet and INFLATE the aft one, so that the latter, becoming heavier, would lower the stern and raise the bow of the vessel.

Along each side of the envelope are seen strips to which the car suspension–cords are attached. To prevent these cords being jerked asunder, by the rolling or pitching of the vessel, horizontal fins, each 172 square feet in area, are provided at each side of the rear end of the balloon. In the past several serious accidents have been caused by the violent pitching of the balloon when caught in a gale, and so severe have been the stresses on the suspension cords that great damage has been done to the envelope, and the aeronauts have been fortunate if they have been able to make a safe descent.

The propeller and engine are carried by the car, which is slung well below the balloon, and by an ingenious contrivance the car always remains in a horizontal position, however much the balloon may be inclined. It is no uncommon occurrence for the balloon to make a considerable angle with the car beneath.

The propeller is quite a work of art. It has a diameter of about 14 feet, and consists of a frame of hollow steel tubes covered with fabric. It is so arranged that when out of action its blades fall lengthwise upon the frame supporting it, but when it is set to work the blades at once open out. The engine weighs 770 pounds, and has six cylinders, which develop 100 horse–power at 1200 revolutions a minute.

The vessel may be steered either to the right or the left by means of a large vertical helm, some 80 square feet in area, which is hinged at the rear end to a fixed vertical plane of 200 square feet area.

An upward or downward inclination is, as we have seen, effected by the ballonets, but in cases of emergency these compensators cannot be deflated or inflated sufficiently rapidly, and a large movable weight is employed for altering the balance of the vessel.

In this country the authorities have hitherto favoured the non–rigid air–ship for military and naval use. The Astra–Torres belongs to this type of vessel, which can be rapidly deflated and transported, and so, too, the air–ship built by Mr. Willows.

CHAPTER XIII. The Zeppelin and Gotha Raids

In the House of Commons recently Mr. Bonar Law announced that since the commencement of the war 14,250 lives had been lost as the result of enemy action by submarines and air–craft. A large percentage of these figures represents women, children, and defenceless citizens.

One had become almost hardened to the German method of making war on the civil population—that system of striving to act upon civilian "nerves" by calculated brutality which is summed up in the word "frightfulness". But the publication of these figures awoke some of the old horror of German warfare. The sum total of lives lost brought home to the people at home the fact that bombardment from air and sea, while it had failed to shake their MORAL, had taken a large toll of human life.

At first the Zeppelin raids were not taken very seriously in this country. People rushed out of their houses to see the unwonted spectacle of an air–ship dealing death and destruction from the clouds. But soon the novelty began to wear off, and as the raids became more frequent and the casualty lists grew larger, people began to murmur against the policy of taking these attacks "lying down". It was felt that "darkness and composure" formed but a feeble and ignoble weapon of defence. The people spoke with no uncertain voice, and it began to dawn upon the authorities that the system of regarding London and the south–east coast as part of "the front" was no excuse for not taking protective measures.

It was the raid into the Midlands on the night of 31st January, 1916, that finally shelved the old policy of do nothing. Further justification, if any were needed, for active measures was supplied by a still more audacious raid upon the east coast of Scotland, upon which occasion Zeppelins soared over England—at their will. Then the authorities woke up, and an extensive scheme of anti–aircraft guns and squadrons of aeroplanes was devised.

About March of the year 1916 the Germans began to break the monotony of the Zeppelin raids by using sea–planes as variants. So there was plenty of work for our new defensive air force. Indeed, people began to ask themselves why we should not hit back by making raids into Germany. The subject was well aired in the public press, and distinguished advocates came forward for and against the policy of reprisals. At a considerably later date reprisals carried the day, and, as we write, air raids by the British into Germany are of frequent occurrence.

In March, 1916, the fruits of the new policy began to appear, and people found them very refreshing. A fleet of Zeppelins found, on approaching the mouth of the Thames, a very warm reception. Powerful searchlights, and shells from new anti–aircraft guns, played all round them. At length a shot got home. One of the Zeppelins, "winged" by a shell, began a wobbly retreat which ended in the waters of the estuary. The navy finished the business. The wrecked air–ship was quickly surrounded by a little fleet of destroyers and patrol–boats, and the crew were brought ashore, prisoners. That same night yet another Zeppelin was hit and damaged in another part of the country.

Raids followed in such quick succession as to be almost of nightly occurrence during the favouring moonless nights. Later, the conditions were reversed, and the attacks by aeroplane were all made in bright moonlight. But ever the defence became more strenuous. Then aeroplanes began to play the role of "hornets", as Mr. Winston Churchill, speaking rather too previously, designated them.

Lieutenant Brandon, R.F.C., succeeded in dropping several aerial bombs on a Zeppelin during the raid on March 31, but it was not until six months later that an airman succeeded in bringing down a Zeppelin on British soil. The credit of repeating Lieutenant Warneford's great feat belongs to Lieutenant W. R. Robinson, and the fight was witnessed by a large gathering. It occurred in the very formidable air raid on the night of September 2. Breathlessly the spectators watched the Zeppelin harried by searchlight and shell–fire. Suddenly it disappeared behind a veil of smoke which it had thrown out to baffle its pursuers. Then it appeared again, and a loud shout went up from the watching thousands. It was silhouetted against the night clouds in a faint line of fire. The hue deepened, the glow spread all round, and the doomed airship began its crash to earth in a smother of flame. The witnesses to this amazing spectacle naturally supposed that a shell had struck the Zeppelin. Its tiny assailant that had dealt the death–blow had been quite invisible during the fight. Only on the following morning did the public learn of

Lieutenant Robinson's feat. It appeared that he had been in the air a couple of hours, engaged in other conflicts with his monster foes. Besides the V.C. the plucky airman won considerable money prizes from citizens for destroying the first Zeppelin on British soil.

The Zeppelin raids continued at varying intervals for the remainder of the year. As the power of the defence increased the air–ships were forced to greater altitudes, with a corresponding decrease in the accuracy with which they could aim bombs on specified objects. But, however futile the raids, and however widely they missed their mark, there was no falling off in the outrageous claims made in the German communiques. Bombs dropped in fields, waste lands, and even the sea, masqueraded in the reports as missiles which had sunk ships in harbour, destroyed docks, and started fires in important military areas. So persistent were these exaggerations that it became evident that the Zeppelin raids were intended quite as much for moral effect at home as for material damage abroad. The heartening effect of the raids upon the German populace is evidenced by the mental attitude of men made prisoners on any of the fronts. Only with the utmost difficulty were their captors able to persuade them that London and other large towns were not in ruins; that shipbuilding was not at a standstill; and that the British people was not ready at any moment to purchase indemnity from the raids by concluding a German peace. When one method of terrorism fails try another, was evidently the German motto. After the Zeppelin the Gotha, and after that the submarine.

The next year—1917—brought in a very welcome change in the situation. One Zeppelin after another met with its just deserts, the British navy in particular scoring heavily against them. Nor must the skill and enterprise of our French allies be forgotten. In March, 1917, they shot down a Zeppelin at Compiegne, and seven months later dealt the blow which finally rid these islands of the Zeppelin menace.

For nearly a year London, owing to its greatly increased defences, had been free from attack. Then, on the night of October 19, Germany made a colossal effort to make good their boast of laying London in ruins. A fleet of eleven Zeppelins came over, five of which found the city. One, drifting low and silently, was responsible for most of the casualties, which totalled 34 killed and 56 injured.

The fleet got away from these shores without mishap. Then, at long last, came retribution. Flying very high, they seem to have encountered an aerial storm which drove them helplessly over French territory. Our allies were swift to seize this golden opportunity.

Their airmen and anti–aircraft guns shot down no less than four of the Zeppelins in broad daylight, one of which was captured whole. Of the remainder, one at least drifted over the Mediterranean, and was not heard of again. That was the last of the Zeppelin, so far as the civilian population was concerned. But, for nearly a year, the work of killing citizens had been undertaken by the big bomb–dropping Gotha aeroplanes.

The work of the Gotha belongs rightly to the second part of this book, which deals with aeroplanes and airmen; but it would be convenient to dispose here of the part played by the Gotha in the air raids upon this country.

The reconnaissance took place on Tuesday, November 28, 1916, when in a slight haze a German aeroplane suddenly appeared over London, dropped six bombs, and flew off. The Gotha was intercepted off Dunkirk by the French, and brought down. Pilot and observer–two naval lieutenants–were found to have a large–scale map of London in their possession. The new era of raids had commenced.

Very soon it became evident that the new squadron of Gothas were much more destructive than the former fleets of unwieldy Zeppelins. These great Gothas were each capable of dropping nearly a ton of bombs. And their heavy armament and swift flight rendered them far less vulnerable than the air–ship.

From March 1 to October 31, 1917, no less than twenty–two raids took place, chiefly on London and towns on the south–east coast. The casualties amounted to 484 killed and 410 wounded. The two worst raids occurred June 13 on East London, and September 3 on the Sheerness and Chatham area.

A squadron of fifteen aeroplanes carried out the raid, on June 13, and although they were only over the city for a period of fifteen minutes the casualty list was exceedingly heavy—104 killed and 432 wounded. Many children were among the killed and injured as the result of a bomb which fell upon a Council school. The raid was carried out in daylight, and the bombs began to drop before any warning could be given. Later, an effective and comprehensive system of warnings was devised, and when people had acquired the habit of taking shelter, instead of rushing out into the street to see the aerial combats, the casualties began to diminish.

It is worthy of record that the possible danger to schools had been anticipated, and for some weeks previously the children had taken part in "Air Raid Drill". When the raid came, the children behaved in the most exemplary fashion. They went through the manoeuvres as though it was merely a rehearsal, and their bearing as well as the coolness of the teachers obviated all danger from panic. In this raid the enemy first made use of aerial torpedoes.

Large loss of life, due to a building being struck, was also the feature of the moonlight raid on September 4. On this occasion enemy airmen found a mark on the Royal Naval barracks at Sheerness. The barracks were fitted with hammocks for sleeping, and no less than 108 bluejackets lost their lives, the number of wounded amounting to 92. Although the raid lasted nearly an hour and powerful searchlights were brought into play, neither guns nor our airmen succeeded in causing any loss to the raiders. Bombs were dropped at a number of other places, including Margate and Southend, but without result.

No less than six raids took place on London before the end of the month, but the greatest number of killed in any one of the raids was eleven, while on September 28 the raiders were driven off before they could claim any victims. The establishment of a close barrage of aerial guns did much to discourage the raiders, and gradually London, from being the most vulnerable spot in the British Isles, began to enjoy comparative immunity from attack.

Paris, too, during the Great War has had to suffer bombardment from the air, but not nearly to the same extent as London. The comparative immunity of Paris from air raids is due partly to the prompt measures which were taken to defend the capital. The French did not wait, as did the British, until the populace was goaded to the last point of exasperation, but quickly instituted the barrage system, in which we afterwards followed their lead. Moreover, the French were much more prompt in adopting retaliatory tactics. They hit back without having to wade through long moral and philosophical disquisitions upon the ethics of "reprisals". On the other hand, it must be remembered that Paris, from the aerial standpoint, is a much more difficult objective than London. The enemy airman has to cross the French lines, which, like his own, stretch for miles in the rear. Practically he is in hostile country all the time, and he has to get back across the same dangerous air zones. It is a far easier task to dodge a few sea-planes over the wide seas en route to London. And on reaching the coast the airman has to evade or fight scattered local defences, instead of penetrating the close barriers which confront him all the way to

Paris.

Since the first Zeppelin attack on Paris on March 21, 1915, when two of the air–ships reached the suburbs, killing 23 persons and injuring 30, there have been many raids and attempted raids, but mostly by single machines. The first air raid in force upon the French capital took place on January 31, 1918, when a squadron of Gothas crossed the lines north of Compiegne. Two hospitals were hit, and the casualties from the raid amounted to 20 killed and 50 wounded.

After the Italian set–back in the winter of 1917, the Venetian plain lay open to aerial bombardment by the Germans, who had given substantial military aid to their Austrian allies. This was an opportunity not to be lost by Germany, and Venice and other towns of the plain were subject to systematic bombardment.

At the time of writing, Germany is beginning to suffer some of the annoyances she is so ready to inflict upon others. The recently constituted Air Ministry have just published figures relating to the air raids into Germany from December 1, 1917, to February 19, 1918 inclusive. During these eleven weeks no fewer than thirty–five raids have taken place upon a variety of towns, railways, works, and barracks. In the list figure such important towns as Mannheim (pop. 20,000) and Metz (pop. 100,000). The average weight of bombs dropped at each raid works out about 1000 lbs. This welcome official report is but one of many signs which point the way to the growing supremacy of the Allies in the air.

PART II. AEROPLANES AND AIRMEN

CHAPTER XIV. Early Attempts in Aviation

The desire to fly is no new growth in humanity. For countless years men have longed to emulate the birds—"To soar upward and glide, free as a bird, over smiling fields, leafy woods, and mirror–like lakes," as a great pioneer of aviation said. Great scholars and thinkers of old, such as Horace, Homer, Pindar, Tasso, and all the glorious line, dreamt of flight, but it has been left for the present century to see those dreams fulfilled.

Early writers of the fourth century saw the possibility of aerial navigation, but those who tried to put their theories in practice were beset by so many difficulties that they rarely succeeded in leaving the ground.

Most of the early pioneers of aviation believed that if a man wanted to fly he must provide himself with a pair of wings similar to those of a large bird. The story goes that a certain abbot told King James IV of Scotland that he would fly from Stirling Castle to Paris. He made for himself powerful wings of eagles' feathers, which he fixed to his body and launched himself into the air. As might be expected, he fell and broke his legs.

But although the muscles of man are of insufficient strength to bear him in the air, it has been found possible, by using a motor engine, to give to man the power of flight which his natural weakness denied him.

Scientists estimate that to raise a man of about 12 stone in the air and enable him to fly there would be required an immense pair of wings over 20 feet in span. In comparison with the weight of a man a bird's weight is remarkably small—the largest bird does not weigh much more than 20 pounds—but its wing muscles are infinitely stronger in proportion than the shoulder and arm muscles of a man.

As we shall see in a succeeding chapter, the "wing" theory was persevered with for many years some two or three centuries ago, and later on it was of much use in providing data for the gradual development of the modern aeroplane.

CHAPTER XV. A Pioneer in Aviation

Hitherto we have traced the gradual development of the balloon right from the early days of aeronautics, when the brothers Montgolfier constructed their hot–air balloon, down to the most modern dirigible. It is now our purpose, in this and subsequent chapters, to follow the course of the pioneers of aviation.

It must not be supposed that the invention of the steerable balloon was greatly in advance of that of the heavier–than–air machine. Indeed, developments in both the dirigible airship and the aeroplane have taken place side by side. In some cases men like Santos Dumont have given earnest attention to both forms of air–craft, and produced practical results with both. Thus, after the famous Brazilian aeronaut had won the Deutsch prize for a flight in an air–ship round the Eiffel tower, he immediately set to work to construct an aeroplane which he subsequently piloted at Bagatelle and was awarded the first "Deutsch prize" for aviation.

It is generally agreed that the undoubted inventor of the aeroplane, practically in the form in which it now appears, was an English engineer, Sir George Cayley. Just over a hundred years ago this clever Englishman worked out complete plans for an aeroplane, which in many vital respects embodied the principal parts of the monoplane as it exists to–day.

There were wings which were inclined so that they formed a lifting plane; moreover, the wings were curved, or "cambered", similar to the wing of a bird, and, as we shall see in a later chapter, this curve is one of the salient features of the plane of a modern heavier–than–air machine. Sir George also advocated the screw propeller worked by some form of "explosion" motor, which at that time had not arrived. Indeed, if there had been a motor available it is quite possible that England would have led the way in aviation. But, unfortunately, owing to the absence of a powerful motor engine, Sir George's ideas could not be practically carried out till nearly a century later, and then Englishmen were forestalled by the Wright brothers, of America, as well as by several French inventors.

The distinguished French writer, Alphonse Berget, in his book, The Conquest of the Air, pays a striking tribute to our English inventor, and this, coming from a gentleman who is writing from a French point of view, makes the praise of great value. In alluding to Sir George, M. Berget says: "The inventor, the incontestable forerunner of aviation, was an Englishman, Sir George Cayley, and it was in 1809 that he described his project in detail in Nicholson's Journal. . . . His idea embodied 'everything'—the wings forming an oblique sail, the empennage, the spindle forms to diminish resistance, the screw–propeller, the 'explosion' motor, . . . he even described a means of securing automatic stability. Is not all that marvellous, and does it not constitute a complete specification for everything in aviation?

"Thus it is necessary to inscribe the name of Sir George Cayley in letters of gold, in the first page of the aeroplane's history. Besides, the learned Englishman did not confine himself to 'drawing–paper': he built the first apparatus (without a motor) which gave him results highly promising. Then he built a second machine, this time with a motor, but unfortunately during the trials it was smashed to pieces."

But were these ideas of any practical value? How is it that he did not succeed in flying, if he had most of the component parts of an aeroplane as we know it to–day?

The answer to the second question is that Sir George did not fly, simply because there was no light petrol motor in existence; the crude motors in use were far too heavy, in proportion to the power developed, for service in a flying machine. It was recognized, not only by Sir George, but by many other English engineers in the first half of the nineteenth century, that as soon as a sufficiently powerful and light engine did appear, then half the battle of the conquest of the air would be won.

But his prophetic voice was of the utmost assistance to such inventors as Santos Dumont, the Wright brothers, M. Bleriot, and others now world–famed. It is quite safe to assume that they gave serious attention to the views held by Sir George, which were given to the world at large in a number of highly–interest– ing lectures and magazine articles. "Ideas" are the very foundation–stones of invention—if we may be allowed the figure of speech—and Englishmen are proud, and rightly proud, to number within their ranks the original inventor of the heavier–than–air machine.

CHAPTER XVI. The "Human Birds"

For many years after the publication of Sir George Cayley's articles and lectures on aviation very little was done in the way of aerial experiments. True, about midway through the nineteenth century two clever engineers, Henson and Stringfellow, built a model aeroplane after the design outlined by Sir George; but though their model was not of much practical value, a little more valuable experience was accumulated which would be of service when the time should come; in other words, when the motor engine should arrive. This model can be seen at the Victoria and Albert Museum, at South Kensington.

A few years later Stringfellow designed a tiny steam–engine, which he fitted to an equally tiny monoplane, and it is said that by its aid he was able to obtain a very short flight through the air. As some recognition of his enterprise the Aeronautical Society, which was founded in 1866, awarded him a prize of L100 for his engine.

The idea of producing a practical form of flying machine was never abandoned entirely. Here and there experiments continued to be carried out, and certain valuable conclusions were arrived at. Many advanced thinkers and writers of half a century ago set forth their opinions on the possibilities of human flight. Some of them, like Emerson, not only believed that flight would come, but also stated why it had not arrived. Thus Emerson, when writing on the subject of air navigation about fifty years ago, remarked: "We think the population is not yet quite fit for them, and therefore there will be none. Our friend suggests so many inconveniences from piracy out of the high air to orchards and lone houses, and also to high fliers, and the total inadequacy of the present system of defence, that we have not the heart to break the sleep of the great public by the repetition of these details. When children come into the library we put the inkstand and the watch on the high shelf until they be a little older."

About the year 1870 a young German engineer, named Otto Lilienthal, began some experiments with a motorless glider, which in course of time were to make him world–famed. For nearly twenty years Lilienthal carried on his aerial research work in secrecy, and it was not until about the year 1890 that his experimental work was sufficiently advanced for him to give demonstrations in public.

The young German was a firm believer in what was known as the "soaring–plane" theory of flight. From the picture here given we can get some idea of his curious machine. It consisted of large wings, formed of thin osiers, over which was stretched light fabric. At the back were two horizontal rudders shaped somewhat like the long forked tail of a swallow, and over these was a large steering rudder. The wings were arranged around the glider's body. The whole apparatus weighed about 40 pounds.

Lilienthal's flights, or glides, were made from the top of a specially–constructed large mound, and in some cases from the summit of a low tower. The "birdman" would stand on the top of the mound, full to the wind, and run quickly forward with outstretched wings. When he thought he had gained sufficient momentum he jumped into the air, and the wings of the glider bore him through the air to the base of the mound.

To preserve the balance of his machine—always a most difficult feat—he swung his legs and hips to one side or the other, as occasion required, and, after hundreds of glides had been made, he became so skilful in maintaining the equilibrium of his machine that he was able to cover a distance, downhill, of 300 yards.

Later on, Lilienthal abandoned the glider, or elementary form of monoplane, and adopted a system of superposed planes, corresponding to the modern biplane. The promising career of this clever German was brought to an untimely end in 1896, when, in attempting to glide from a height of about 80 yards, his apparatus made a sudden downward swoop, and he broke his neck.

Now that Lillenthal's experiments had proved conclusively the efficiency of wings, or planes, as carrying surfaces, other engineers followed in his footsteps, and tried to improve on his good work.

The first "birdman" to use a glider in this country was Mr. Percy Pilcher who carried out his experiments at Cardross in Scotland. His glides were at first made with a form of apparatus very similar to that employed by Lilienthal, and in time he came to use much larger machines. So cumbersome, however, was his apparatus—it weighed nearly 4 stones—that with such a great weight upon his shoulders he could not run forward quickly enough to gain sufficient momentum to "carry off" from the hillside. To assist him in launching the apparatus the machine was towed by horses, and when sufficient impetus had been gained the tow–rope was cast off.

Three years after Lilienthal's death Pilcher met with a similar accident. While making a flight his glider was overturned, and the unfortunate "birdman " was dashed to death.

In America there were at this time two or three "human birds", one of the most famous being M. Octave Chanute. During the years 1895–7 Chanute made many flights in various types of gliding machines, some of which had as many as half a dozen planes arranged one above another. His best results, however, were obtained by the two–plane machine, resembling to a remarkable extent the modern biplane.

CHAPTER XVII. The Aeroplane and the Bird

We have seen that the inventors of flying machines in the early days of aviation modelled their various craft somewhat in the form of a bird, and that many of them believed that if the conquest of the air was to be achieved man must copy nature and provide himself with wings.

Let us closely examine a modern monoplane and discover in what way it resembles the body of a bird in build.

First, there is the long and comparatively narrow body, or FUSELAGE, at the end of which is the rudder, corresponding to the bird's tail. The chassis, or under carriage, consisting of wheels, skids, may well be compared with the legs of a bird, and the planes are very similar in construction to the bird's wings. But here the resemblance ends: the aeroplane does not fly, nor will it ever fly, as a bird flies.

If we carefully inspect the wing of a bird—say a large bird, such as the crow—we shall find it curved or arched from front to back. This curve, however, is somewhat irregular. At the front edge of the wing it is sharpest, and there is a gradual dip or slope backwards and downwards. There is a special reason for this peculiar structure, as we shall see in a later chapter.

Now it is quite evident that the inventors of aeroplanes have modelled the planes of their craft on the bird's wing. Strictly speaking, the word "plane" is a misnomer when applied to the supporting structure of an aeroplane. Euclid defines a plane, or a plane surface, as one in which, any two points being taken, the straight line between them lies wholly in that surface. But the plane of a flying machine is curved, or CAMBERED, and if one point were taken on the front of the so–called plane, and another on the back, a straight line joining these two points could not possibly lie wholly on the surface.

All planes are not cambered to the same extent: some have a very small curvature; in others the curve is greatly pronounced. Planes of the former type are generally fitted to racing aeroplanes, because they offer less resistance to the air than do deeply–cambered planes. Indeed, it is in the degree of camber that the various types of flying machine show their chief diversity, just as the work of certain shipmasters is known by the particular lines of the bow and stern of the vessels which are built in their yards.

Birds fly by a flapping movement of their wings, or by soaring. We are quite familiar with both these actions: at one time the bird propels itself by means of powerful muscles attached to its wings by means of which the wings are flapped up and down; at another time the bird, with wings nicely adjusted so as to take advantage of all the peculiarities of the air currents, keeps them almost stationary, and soars or glides through the air.

The method of soaring alone has long since been proved to be impracticable as a means of carrying a machine through the air, unless, of course, one describes the natural glide of an aeroplane from a great height down to earth as soaring. But the flapping motion was not proved a failure until numerous experiments by early aviators had been tried.

Probably the most successful attempt at propulsion by this method was that of a French locksmith named Besnier. Over two hundred years ago he made for himself a pair of light wooden paddles, with blades at either end, somewhat similar in shape to the double paddle of a canoe. These he placed over his shoulders, his feet being attached by ropes to the hindmost paddles. Jumping off from some high place in the face of a stiff breeze, he violently worked his arms and legs, so that the paddles beat the air and gave him support. It is said that Besnier became so expert in the management of his simple apparatus that he was able to raise himself from the ground, and skim lightly over fields and rivers for a considerable distance.

Now it has been shown that the enormous extent of wing required to support a man of average weight would be much too large to be flapped by man's arm muscles. But in this, as with everything else, we have succeeded in harnessing the forces of nature into our service as tools and machinery.

And is not this, after all, one of the chief, distinctions between man and the lower orders of creation? The latter fulfil most of their bodily requirements by muscular effort. If a horse wants to get from one place to another it walks; man can go on wheels. None of the lower animals makes a single tool to assist it in the various means of sustaining life; but man puts on his "thinking–cap", and invents useful machines and tools to enable him to assist or dispense with muscular movement.

Thus we find that in aviation man has designed the propeller, which, by its rapid revolutions derived from the motive power of the aerial engine, cuts a spiral pathway through the air and drives the light craft rapidly forward. The chief use of the planes is for support to the machine, and the chief duty of the pilot is to balance and steer the craft by the manipulation of the rudder, elevation and warping controls.

CHAPTER XVIII. A Great British Inventor of Aeroplanes

Though, as we have seen, most of the early attempts at aerial navigation were made by foreign engineers, yet we are proud to number among the ranks of the early inventors of heavier–than–air machines Sir Hiram Maxim, who, though an American by birth, has spent most of his life in Britain and may therefore be called a British inventor.

Perhaps to most of us this inventor's name is known more in connection with the famous "Maxim" gun, which he designed, and which was named after him. But as early as 1894, when the construction of aeroplanes was in a very backward state, Sir Hiram succeeded in making an interesting and ingenious aeroplane, which he proposed to drive by a particularly light steam–engine.

Sir Hiram's first machine, which was made in 1890, was designed to be guided by a double set of rails, one set arranged below and the other above its running wheels. The intention was to make the machine raise itself just off the ground rails, but yet be prevented from soaring by the set of guard rails above the wheels, which acted as a check on it. The motive force was given by a very powerful steam–engine of over 300 horse–power, and this drove two enormous propellers, some 17 feet in length. The total weight of the machine was 8000 pounds, but even with this enormous weight the engine was capable of raising the machine from the ground.

For three or four years Sir Hiram made numerous experiments with his aeroplane, but in 1894 it broke through the upper guard rail and turned itself over among the surrounding trees, wrecking itself badly.

But though the Maxim aeroplane did not yield very practical results, it proved that if a lighter but more powerful engine could be made, the chief difficulty iii the way of aerial flight would be removed. This was soon forthcoming in the invention of the petrol motor. In a lecture to the Scottish Aeronautical Society, delivered in Glasgow in November, 1913, Sir Hiram claimed to be the inventor of the first machine which actually rose from the earth. Before the distinguished inventor spoke of his own work in aviation he recalled experiments made by his father in 1856–7, when Sir Hiram was sixteen years of age. The flying machine designed by the elder Maxim consisted of a small platform, which it was proposed to lift directly into the air by the action of two screw–propellers revolving in reverse directions. For a motor the inventor intended to employ some kind of explosive material, gunpowder preferred, but the lecturer distinctly remembered that his father said that if an apparatus could be successfully navigated through the air it would be of such inevitable value as a military engine that no matter how much it might cost to run it would be used by Governments.

Of his own claim as an inventor of air–craft it would be well to quote Sir Hiram's actual words, as given by the Glasgow Herald, which contained a full report of the lecture.

"Some forty years ago, when I commenced to think of the subject, my first idea was to lift my machint by vertical propellers, and I actually commenced drawings and made calculations for a machine on that plan, using an oil motor, or something like a Brayton engine, for motive power. However, I was completely unable to work out any system which would not be too heavy to lift itself directly into the air, and it was only when I commenced to study the aeroplane system that it became apparent to me that it would be possible to make a machine light enough and powerful enough to raise itself without the agency of a balloon. From the first I was convinced that it would be quite out of the question to employ a balloon in any form. At that time the light high–speed petrol motor had no existence. The only power available being steam–engines, I made all my calculations with a view of using steam as the motive power. While I was studying the question of the possibility of making a flying machine that would actually fly, I became convinced that there was but one system to work on, and that was the aeroplane system. I made many calculations, and found that an aeroplane machine driven by a steam–engine ought to lift itself into the air."

Sir Hiram then went on to say that it was the work of making an automatic gun which was the direct cause of his experiments with flying machines. To continue the report:

"One day I was approached by three gentle– men who were interested in the gun, and they asked me if it would be possible for me to build a flying machine, how long it would take, and how much it would cost. My reply was that it would take five years and would cost L50,000. The first three years would be devoted to developing a light internal–combustion engine, and the remaining two years to making a flying machine.

"Later on a considerable sum of money was placed at my disposal, and the experiments commenced, but unfortunately the gun business called for my attention abroad, and during the first two years of the experimental work I was out of England eighteen months.

"Although I had thought much of the internal–combustion engine it seemed to me that it would take too long to develop one and that it would be a hopeless task in my absence from England; so I decided that in my first experiments at least I would use a steam–engine. I therefore designed and made a steam–engine and boiler of which Mr. Charles Parsons has since said that, next to the Maxim gun, it developed more energy for its weight than any other heat engine ever made. That was true at the time, but is very wide of the mark now."

Speaking of motors, the veteran lecturer remarked: "Perhaps there was no problem in the world on which mathematicians had differed so widely as on the problem of flight. Twenty years ago experimenters said: 'Give us a motor that will develop 1 horse–power with the weight of a barnyard fowl, and we will very soon fly.' At the present moment they had motors which would develop over 2 horse–power and did not weigh more than a 12–pound barnyard fowl. These engines had been developed—I might say created—by the builders of motor cars. Extreme lightness had been gradually obtained by those making racing cars, and that had been intensified by aviators. In many cases a speed of 80 or 100 miles per hour had been attained, and machines had remained in the air for hours and had flown long distances. In some cases nearly a ton had been carried for a short distance."

Such words as these, coming from the lips of a great inventor, give us a deep insight into the working of the inventor's mind, and, incidentally, show us some of the difficulties which beset all pioneers in their tasks. The science of aviation is, indeed, greatly indebted to these early inventors, not the least of whom is the gallant Sir Hiram Maxim.

CHAPTER XIX. The Wright Brothers and their Secret Experiments

In the beginning of the twentieth century many of the leading European newspapers contained brief reports of aerial experiments which were being carried out at Dayton, in the State of Ohio, America. So wonderful were the results of these experiments, and so mysterious were the movements of the two brothers—Orville and Wilbur Wright—who conducted them, that many Europeans would not believe the reports.

No inventors have gone about their work more carefully, methodically, and secretly than did these two Americans, who, hidden from prying eyes, "far from the madding crowd", obtained results which brought them undying fame in the world of aviation.

For years they worked at their self–imposed task of constructing a flying machine which would really soar among the clouds. They had read brief accounts of the experiments carried out by Otto Lilienthal, and in many ways the ground had been well paved for them. It was their great ambition to become real "human birds"; "birds" that would not only glide along down the hillside, but would fly free and unfettered, choosing their aerial paths of travel and their places of destination.

Though there are few reliable accounts of their work in those remote American haunts, during the first six years of the present century, the main facts of their life–history are now well known, and we are able to trace their experiments, step by step, from the time when they constructed their first simple aeroplane down to the appearance of the marvellous biplane which has made them world–famed.

For some time the Wrights experimented with a glider, with which they accomplished even more wonderful results than those obtained by Lilienthal. These two young American engineers—bicyclemakers by trade—were never in a hurry. Step by step they made progress, first with kites, then with small gliders, and ultimately with a large one. The latter was launched into the air by men running forward with it until sufficient momentum had been gained for the craft to go forward on its own account.

The first aeroplane made by the two brothers was a very simple one, as was the method adopted to balance the craft. There were two main planes made of long spreads of canvas arranged one above another, and on the lower plane the pilot lay. A little plane in front of the man was known as the ELEVATOR, and it could be moved up and down by the pilot; when the elevator was tilted up, the aeroplane ascended, when lowered, the machine descended.

At the back was a rudder, also under control of the pilot. The pilot's feet, in a modern aeroplane, rest upon a bar working on a central swivel, and this moves the rudder. To turn to the left, the left foot is moved forward; to turn to the right the right foot.

But it was in the balancing control of their machine that the Wrights showed such great ingenuity. Running from the edges of the lower plane were some wires which met at a point where the pilot could control them. The edges of the plane were flexible; that is, they could be bent slightly either up or down, and this movement of the flexible plane is known as WING WARPING.

You know that when a cyclist is going round a curve his machine leans inwards. Perhaps some of you have seen motor races, such as those held at Brooklands; if so, you must have noticed that the track is banked very steeply at the corners, and when the motorist is going round these corners at, say, 80 miles an hour, his motor makes a considerable angle with the level ground, and looks as if it must topple over. The aeroplane acts in a similar manner, and, unless some means are taken to prevent it, it will turn over.

Let us now see how the pilot worked the "Wright" glider. Suppose the machine tilted down on one side, while in the air, the pilot would pull down, or warp, the edges of the planes on that side of the machine which was the lower. By an ingenious contrivance, when one side was warped down, the other was warped up, with the effect that the machine would be brought back into a horizontal position. (As we shall return to the subject of wing warping in a later chapter, we need not discuss it further here.)

It must not be imagined that as soon as the Wrights had constructed a glider fitted with this clever system of controlling mechanism they could fly when and where they liked. They had to practise for two or three years before they were satisfied with the results of their experiments: neglecting no detail, profiting by their failures, and moving logically from step to step. They never attempted an experiment rashly: there was always a reason for what they did. In fact, their success was due to systematic progress, achieved by wonderful perseverance.

But now, for a short time, we must leave the pioneer work of the Wright brothers, and turn to the invention of the petrol engine as applied to the motor car, an invention which was destined to have far–reaching results on the science of aviation.

CHAPTER XX. The Internal–combustion Engine

We have several times remarked upon the great handicap placed upon the pioneers of aviation by the absence of a light but powerful motor engine. The invention of the internal–combustion engine may be said to have revolutionized the science of flying; had it appeared a century ago, there is no reason to doubt that Sir George Cayley would have produced an aeroplane giving as good results as the machines which have appeared during the last five or six years.

The motor engine and the aeroplane are inseparably connected; one is as necessary to the other as clay is to the potter's wheel, or coal to the blast–furnace. This being the case, it is well that we trace briefly the development of the engine during the last quarter of a century.

The original mechanical genius of the motoring industry was Gottlieb Daimler, the founder of the immense Daimler Motor Works of Coventry. Perhaps nothing in the world of industry has made more rapid strides during the last twenty years than automobilism. In 1900 our road traction was carried on by means of horses; now, especially in the large cities, it is already more than half mechanical, and at the present rate of progress it bids fair to be soon entirely horseless.

About the year 1885 Daimler was experimenting with models of a small motor engine, and the following year he fitted one of his most successful models to a light wagonette. The results were so satisfactory, that in 1888 he took out a patent for an internal–combustion engine—as the motor engine is technically called—and the principle on which this engine was worked aroused great enthusiasm on the Continent.

Soon a young French engineer, named Levassor, began to experiment with models of motor engines, and in 1889 he obtained, with others, the Daimler rights to construct similar engines in France. From now on, French engineers began to give serious attention to the new engine, and soon great improvements were made in it. All this time Britain held aloof from the motor–car; indeed, many Britons scoffed at the idea of mechanically–propelled vehicles, saying that the time and money required for their development would be wasted.

During the years 1888–1900 strange reports of smooth–moving, horseless cars, frequently appearing in public in France, began to reach Britain, and people wondered if the French had stolen a march on us, and if there were anything in the new invention after all. Our engineers had just begun to grasp the immense possibilities of Daimler's engine, but the Government gave them no encouragement.

At length the Hon. Evelyn Ellis, one of the first British motorists, introduced the "horseless carriage" into this country, and the following account of his early trips, which appeared in the Windsor and Eton Express of 27th July, 1895, may be interesting.

"If anyone cares to run over to Datchet, they will see the Hon. Evelyn Ellis, of Rosenau, careering round the roads, up hill and down dale, and without danger to life or limb, in his new motor carriage, which he brought over a short time ago from Paris.

"In appearance it is not unlike a four–wheeled dog–cart, except that the front part has a hood for use on long "driving" tours, in the event of wet weather; it will accommodate four persons, one of whom, on the seat behind, would, of course, be the 'groom', a misnomer, perhaps, for carriage attendant. Under the front seat are receptacles, one for tools with which to repair damages, in the event of a breakdown on the road, and the other for a store of oil, petroleum, or naphtha in cans, from which to replenish the oil tank of the carriage on the journey, if it be a long one.

"Can it be easily driven? We cannot say that such a vehicle would be suitable for a lady, unless rubber–tyred wheels and other improvements are made to the carriage, for a grim grip of the steering handle and a keen eye are necessary for its safe guidance, more especially if the high road be rough. It never requires to be fed, and as it is, moreover, unsusceptible of fatigue, it is obviously the sort of vehicle that should soon achieve a widespread popularity in this country.

"It is a splendid hill climber, and, in fact, such a hill as that of Priest Hill (a pretty good test of its capabilities) shows that it climbs at a faster pace than a pedestrian can walk.

"A trip from Rosenau to Old Windsor, to the entrance of Beaumont College, up Priest Hill, descending the steep, rough, and treacherous hill on the opposite side by Woodside Farm, past the workhouse, through old Windsor, and back to Rosenau within an hour, amply demonstrated how perfectly under control this carriage is, while the sensation of being whirled rapidly along is decidedly pleasing."

Another pioneer of motorism was the Hon. C. S. Rolls, whose untimely death at Bournemouth in 1910, while taking part in the Bournemouth aviation meeting, was deeply deplored all over the country. Mr. Rolls made a tour of the country in a motor–car in 1895, with the double object of impressing people with the stupidity of the law with regard to locomotion, and of illustrating the practical possibilities of the motor. You may know that Mr. Rolls was the first man to fly across the Channel, and back again to Dover, without once alighting.

CHAPTER XXI. The Internal–combustion Engine(Cont.)

I suppose many of my readers are quite familiar with the working of a steam–engine. Probably you have owned models of steam–engines right from your earliest youth, and there are few boys who do not know how the railway engine works.

But though you may be quite familiar with the mechanism of this engine, it does not follow that you know how the petrol engine works, for the two are highly dissimilar. It is well, therefore, that we include a short description of the internal–combustion engine such as is applied to motor–cars, for then we shall be able to understand the principles of the aeroplane engine.

At present petrol is the chief fuel used for the motor engine. Numerous experiments have been tried with other fuels, such as benzine, but petrol yields the best results.

Petrol is distilled from oil which comes from wells bored deep down in the ground in Pennsylvania, in the south of Russia, in Burma, and elsewhere. Also it is distilled in Scotland from oil shale, from which paraffin oil and wax and similar substances are produced. When the oil is brought to the surface it contains many impurities, and in its native form is unsuitable for motor engines. The crude oil is composed of a number of different kinds of oil; some being light and clear, others heavy and thick.

To purify the oil it is placed in a large metal vessel or "still". Steam is first passed over the oil in the still, and this changes the lightest of the oils into vapours. These vapours are sent through a series of pipes surrounded with cold water, where they are cooled and become liquid again. Petrol is a mixture of these lighter products of the oil.

If petrol be placed in the air it readily turns into a vapour, and this vapour is extremely inflammable. For this reason petrol is always kept in sealed tins, and very large quantities are not allowed to be stored near large towns. The greatest care has to be exercised in the use of this "unsafe" spirit. For example, it is most dangerous to smoke when filling a tank with petrol, or to use the spirit near a naked light. Many motor–cars have been set on fire through the petrol leaking out of the tank in which it is carried.

The tank which contains the petrol is placed under one of the seats of the motor–car, or at the rear; if in use on a motor–cycle it is arranged along the top bar of the frame, just in front of the driver. This tank is connected to the "carburettor", a little vessel having a small nozzle projecting upwards in its centre. The petrol trickles from the tank into the carburettor, and is kept at a constant level by means of a float which acts in a very similar way to the ballcock of a water cistern.

The carburettor is connected to the cylinder of the engine by another pipe, and there is valve which is opened by the engine itself and is closed by a spring. By an ingenious contrivance the valve is opened when the piston moves out of the cylinder, and a vacuum is created behind it and in the carburettor. This carries a fine spray of petrol to be sucked up through the nozzle. Air is also sucked into the carburettor, and the mixture of air and petrol spray produces an inflammable vapour which is drawn straight into the cylinder of the engine.

As soon as the piston moves back, the inlet valve is automatically closed and the vapour is compressed into the top of the cylinder. This is exploded by an electric spatk, which is passed between two points inside the cylinder, and the force of the explosion drives the piston outwards again. On its return the "exhaust" or burnt gases are driven out through another valve, known as the "exhaust" valve.

Whether the engine has two, four, or six cylinders, the car is propelled in a similar way for all the pistons assist in turning one shaft, called the engine shaft, which runs along the centre of the car to the back axle.

The rapid explosions in the cylinder produce great heat, and the cylinders are kept cool by circulating water round them. When the water has become very hot it passes through a number of pipes, called the "radiator", placed in front of the car; the cold air rushing between the coils cools the water, so that it can be used over and over again.

No water is needed for the engine of a motor cycle. You will notice that the cylinders are enclosed by wide rings of metal, and these rings are quite sufficient to radiate the heat as quickly as it is generated.

CHAPTER XXII. The Aeroplane Engine

We have seen that a very important part of the internal–combustion engine, as used on the motor–car, is the radiator, which prevents the engine from becoming overheated and thus ceasing to work. The higher the speed at which the engine runs the hotter does it become, and the greater the necessity for an efficient cooling apparatus.

But the motor on an aeroplane has to do much harder work than the motor used for driving the motor–car, while it maintains a much higher speed. Thus there is an even greater tendency for it to become overheated; and the great problem which inventors of aeroplane engines have had to face is the construction of a light but powerful engine equipped with some apparatus for keeping it cool.

Many different forms of aeroplane engines have been invented during the last few years. Some inventors preferred the radiator system of cooling the engine, but the tank containing the water, and the radiator itself, added considerably to the weight of the motor, and this, of course, was a serious drawback to its employment.

But in 1909 there appeared a most ingeniously–constructed engine which was destined to take a very prominent part in the progress of aviation. This was the famous "Gnome" engine, by means of which races almost innumerable have been won, and amazing records established.

We have already referred to the engine shaft of the motor–car, which is revolved by the pistons of the various fixed cylinders. In all aeroplane engines which had appeared before the Gnome the same principle of construction had been adopted; that is to say, the cylinders were fixed, and the engine shaft revolved.

But in the Gnome engine the reverse order of things takes place; the shaft is fixed, and the cylinders fly round it at a tremendous speed. Thus the rapid whirl in the air keeps the engine cool, and cumbersome tanks and unwieldy radiators can be dispensed with. This arrangement enabled the engine to be made very light and yet be of greater horse–power than that attained by previously–existing engines.

A further very important characteristic of the rotary–cylinder engine is that no flywheel is used; in a stationary engine it has been found necessary to have a fly–wheel in addition to the propeller. The rotary–cylinder engine acts as its own fly–wheel, thus again saving considerable weight.

The new engine astonished experts when they first examined it, and all sorts of disasters to it were predicted. It was of such revolutionary design that wiseacres shook their heads and said that any pilot who used it would be constantly in trouble with it. But during the last few years it has passed from one triumph to another, commencing with a long–distance record established by Henri Farman at Rheims, in 1909. It has since been used with success by aviators all the world over. That in the Aerial Derby of 1913—which was flown over a course Of 94 miles around London—six of the eleven machines which took part in the race were fitted with Gnome engines, and victory was achieved by Mr. Gustav Hamel, who drove an 80–horse–power Gnome, is conclusive evidence of the high value of this engine in aviation.

CHAPTER XXIII. A Famous British Inventor of Aviation Engines

In the general design and beauty of workmanship involved in the construction of aeroplanes, Britain is now quite the equal of her foreign rivals; even in engines we are making extremely rapid progress, and the well–known Green Engine Company, profiting by the result of nine years' experience, are able to turn out aeroplane engines as reliable, efficient, and as light in pounds weight per horse–power as any aero engine in existence.

In the early days of aviation larger and better engines of British make specially suited for aeroplanes were our most urgent need.

The story of the invention of the "Green" engine is a record of triumph over great difficulties.

Early in 1909—the memorable year when M. Bleriot was firing the enthusiasm of most engineers by his cross–Channel flight; when records were being established at Rheims; and when M. Paulhan won the great prize of L10,000 for the London to Manchester flight— Mr. Green conceived a number of ingenious ideas for an aero engine.

One of Mr. Green's requirements was that the cylinders should be made of cast–steel, and that they should come from a British foundry. The company that took the work in hand, the Aster Company, had confidence in the inventor's ideas. It is said that they had to waste 250 castings before six perfect cylinders were produced. It is estimated that the first Green engine cost L6000. These engines can be purchased for less than L500.

The closing months of 1909 saw the Green engine firmly established. In October of that year Mr. Moore Brabazon won the first all–British competition of L1000 offered by the Daily Mail for the first machine to fly a circular mile course. His aeroplane was fitted with a 60–horse–power Green aero engine. In the same year M. Michelin offered L1000 for a long–distance flight in all–British aviation; this prize was also won by Mr. Brabazon, who made a flight of 17 miles.

Some of Colonel Cody's achievements in aviation were made with the Green engine. In 1910 he succeeded in winning both the duration and cross–country Michelin competitions, and in 1911 he again accomplished similar feats. In this year he also finished fourth in the all–round–Britain race. This was a most meritorious performance when it is remembered that his Cathedral weighed nearly a ton and ahalf, and that the 60–horse–power Green was practically "untouched", to use an engineering expression, during the whole of the 1010–mile flight.

The following year saw Cody winning another Michelin prize for a cross–country competition. Here he made a flight of over 200 miles, and his high opinion of the engine may be best described in the letter he wrote to the company, saying: "If you kept the engine supplied from without with petrol and oil, what was within would carry you through".

But the pinnacle of Mr. Green's fame as an inventor was reached in 1913, when Mr. Harry Hawker made his memorable waterplane flight from Cowes to Lough Shinny, an account of which appears in a later chapter. His machine was fitted with a 100–horse–power Green, and with it he flew 1043 miles of the 1540–miles course.

Though the complete course was not covered, neither Mr. Sopwith— who built the machine and bore the expenses of the flight—nor Mr. Hawker attached any blame to the engine. At a dinner of the Aero Club, given in 1914, Mr. Sopwith was most enthusiastic in discussing the merits of the "Green", and after Harry Hawker had recovered from the effects of his fall in Lough Shinny he remarked in reference to the engine: "It is the best I have ever met. I do not know any other that would have done anything like the work."

At the same time that this race was being held the French had a competition from Paris to Deauville, a distance of about 160 miles. When compared with the time and distance covered by Mr. Hawker, the results achieved by the French pilots, flying machines fitted with French engines, were quite insignificant; thus proving how the British industry had caught up, and even passed, its closest rivals.

In 1913 Mr. Grahame White, with one of the 100–horse–power "Greens" succeeded in winning the duration Michelin with a flight of over 300 miles, carrying a mechanic and pilot, 85 gallons of petrol, and 12 gallons of lubricating oil. Compulsory landings were made every 63 miles, and the engine was stopped. In spite of these trying conditions, the engine ran, from start to finish, nearly nine hours without the slightest trouble.

Sufficient has been said to prove conclusively that the thought and labour expended in the perfecting of the Green engine have not been fruitless.

CHAPTER XXIV. The Wright Biplane (Camber of Planes)

Now that the internal–combustion engine had arrived, the Wrights at once commenced the construction of an aeroplane which could be driven by mechanical power. Hitherto, as we have seen, they had made numerous tests with motorless gliders; but though these tests gave them much valuable information concerning the best methods of keeping their craft on an even keel while in the air, they could never hope to make much progress in practical flight until they adopted motor power which would propel the machine through the air.

We may assume that the two brothers had closely studied the engines patented by Daimler and Levassor, and, being of a mechanical turn of mind themselves, they were able to build their own motor, with which they could make experiments in power–driven flight.

Before we study the gradual progress of these experiments it would be well to describe the Wright biplane. The illustration facing p. 96 shows a typical biplane, and though there are certain modifications in most modern machines, the principles upon which it was built apply to all aeroplanes.

The two main supporting planes, A, B, are made of canvas stretched tightly across a light frame, and are slightly curved, or arched, from front to back. This curve is technically known as the CAMBER, and upon the camber depend the strength and speed of the machine.

If you turn back to Chapter XVII you will see that the plane is modelled after the wing of a bird. It has been found that the lifting power of a plane gradually dwindles from the front edge— or ENTERING EDGE, as it is called—backwards. For this reason it is necessary to equip a machine with a very long, narrow plane, rather than with a comparatively broad but short plane.

Perhaps a little example will make this clear. Suppose we had two machines, one of which was fitted with planes 144 feet long and 1 foot wide, and the other with planes 12 feet square. In the former the entering edge of the plane would be twelve times as great as in the latter, and the lifting power would necessarily be much greater. Thus, though both machines have planes of the same area, each plane having a surface of 144 square feet, yet there is a great difference in the "lift" of the two.

But it is not to be concluded that the back portion of a plane is altogether wasted. Numerous experiments have taught aeroplane constructors that if the plane were slightly curved from front to back the rear portion of the plane also exercised a "lift"; thus, instead of the air being simply cut by the entering edge of the plane, it is driven against the arched back of the plane, and helps to lift the machine into the air, and support it when in flight.

There is also a secondary lifting impulse derived from this simple curve. We have seen that the air which has been cut by the front edge of the plane pushes up from below, and is arrested by the top of the arch, but the downward dip of the rear portion of the plane is of service in actually DRAWING THE AIR FROM ABOVE. The rapid air stream which has been cut by the entering edge passes above the top of the curve, and "sucks up", as it were, so that the whole wing is pulled upwards. Thus there are two lifting impulses: one pushing up from below, the other sucking up from above.

It naturally follows that when the camber is very pronounced the machine will fly much slower, but will bear a greater weight than a machine equipped with planes having little or no camber. On high–speed machines, which are used chiefly for racing purposes, the planes have very little camber. This was particularly noticeable in the monoplane piloted by Mr. Hamel in the Aerial Derby of 1913: the wings of this machine seemed to be quite flat, and it was chiefly because of this that the pilot was able to maintain such marvellous speed.

The scientific study of the wing lift of planes has proceeded so far that the actual "lift" can now be measured, providing the speed of the machine is known, together with the superficial area of the planes. The designer can calculate what weight each square foot of the planes will support in the air. Thus some machines have a "lift" of 9 or 10 pounds to each square foot of wing surface, while others are reduced to 3 or 4 pounds per square foot.

CHAPTER XXV. The Wright Biplane (Cont.)

The under part of the frame of the Wright biplane, technically known as the CHASSIS, resembled a pair of long "runner" skates, similar to those used in the Fens for skating races. Upon those runners the machine moved along the ground when starting to fly. In more modern machines the chassis is equipped with two or more small rubber–tyred wheels on which the machine runs along the ground before rising into the air, and on which it alights when a descent is made.

You will notice that the pilot's seat is fixed on the lower plane, and almost in the centre of it, while close by the engine is mounted. Alongside the engine is a radiator which cools the water that has passed round the cylinder of the engine in order to prevent them from becoming overheated.

Above the lower plane is a similar plane arranged parallel to it, and the two are connected by light upright posts of hickory wood known as STRUTS. Such an aeroplane as this, which is equipped with two main planes, known as a BIPLANE. Other types of air–craft are the MONOPLANE, possessing one main plane, and the TRIPLANE, consisting of three planes. No practical machine has been built with more than three main planes; indeed, the triplane is now almost obsolete.

The Wrights fitted their machine with two long–bladed wooden screws, or propellers, which by means of chains and sprocket–wheels, very like those of a bicycle, were driven by the engine, whose speed was about 1200 revolutions a minute. The first motor engine used by these clever pioneers had four cylinders, and developed about 20 horsepower. Nowadays engines are produced which develop more than five times that power.

In later machines one propeller is generally thought to be sufficient; in fact many constructors believe that there is danger in a two–propeller machine, for if one propeller got broken, the other propeller, working at full speed, would probably overturn the machine before the pilot could cut off his engine.

Beyond the propellers there are two little vertical planes which can be moved to one side or the other by a control lever in front of the pilot's seat. These planes or rudders steer the machine from side to side, answering the same purpose as the rudder of a boat.

In front of the supporting planes there are two other horizontal planes, arranged one above the other; these are much smaller than the main planes, and are known as the ELEVATORS. Their function is to raise or lower the machine by catching the air at different angles.

Comparison with a modern biplane, such as may be seen at an aerodrome on any "exhibition" day, will disclose several marked differences in construction between the modern type and the earlier Wright machine, though the central idea is the same.

CHAPTER XXVI. How the Wrights launched their Biplane

Those of us who have seen an aeroplane rise from the ground know that it runs quickly along for 50 or 60 yards, until sufficient momentum has been gained for the craft to lift itself into the air. The Wrights, as stated, fitted their machine with a pair of launching runners which projected from the under side of the lower plane like two very long skates, and the method of launching their craft was quite different from that followed nowadays.

The launching apparatus consisted of a wooden tower at the starting end of the launching ways—a wooden rail about 60 or 70 feet in length. To the top of the tower a weight of about 1/2 ton was suspended. The suspension rope was led downwards over pulleys, thence horizontally to the front end and back to the inner end of the railway, where it was attached to the aeroplane. A small trolley was fitted to the chassis of the machine and this ran along the railway.

To launch the machine, which, of course, stood on the rail, the propellers were set in motion, and the 1/2–ton weight at the top of the tower was released. The falling weight towed the aeroplane rapidly forward along the rail, with a velocity sufficient to cause it to glide smoothly into the air at the other end of the launching ways. By an ingenious arrangement the trolley was left behind on the railway.

It will at once occur to you that there were disadvantages in this system of commencing a flight. One was that the launching apparatus was more or less a fixture. At any rate it could not be carried about from place to place very readily: Supposing the biplane could not return to its starting–point, and the pilot was forced to descend, say, 10 or 12 miles away: in such a case it would be neces– sary to tow the machine back to the launching ways, an obviously inconvenient arrangement, especially in unfavourable country.

For some time the "wheeled" chassis has been in universal use, but in a few cases it has been thought desirable to adopt a combination of runners and wheels. A moderately firm surface is necessary for the machine to run along the ground; if the ground be soft or marly the wheels would sink in the soil, and serious accidents have resulted from the sudden stoppage of the forward motion due to this cause.

With their first power–driven machine the Wrights made a series of very fine flights, at first in a straight line. In 1904 they effected their first turn. By the following year they had made such rapid progress that they were able to exceed a distance of 20 miles in one flight, and keep up in the air for over half an hour at a time. Their manager now gave their experiments great publicity, both in the American and European Press, and in 1908 the brothers, feeling quite sure of their success, emerged from a self–imposed obscurity, and astonished the world with some wonderful flights, both in America and on the French flying ground at Issy.

A great loss to aviation occurred on 30th May, 1912, when Wilbur Wright died from an attack of typhoid fever. His work is officially commemorated in Britain by an annual Premium Lecture, given under the auspices of the Aeronautical Society.

CHAPTER XXVII. The First Man to Fly in Europe

In November, 1906, nearly the whole civilized world was astonished to read that a rich young Brazilian aeronaut, residing in France, had actually succeeded in making a short flight, or, shall we say, an enormous "hop", in a heavier–than–air machine.

This pioneer of aviation was M. Santos Dumont. For five or six years before his experiments with the aeroplane he had made a great many flights in balloons, and also in dirigible balloons. He was the son of well–to–do parents—his father was a successful coffee planter—and he had ample means to carry on his costly experiments.

Flying was Santos Dumont's great hobby. Even in boyhood, when far away in Brazil, he had been keenly interested in the work of Spencer, Green, and other famous aeronauts, and aeronautics became almost a passion with him.

Towards the end of the year 1898 he designed a rather novel form of air–ship. The balloon was shaped like an enormous cigar, some 80 feet long, and it was inflated with about 6000 cubic feet of hydrogen. The most curious contrivance, however, was the motor. This was suspended from the balloon, and was somewhat similar to the small motor used on a motor–cycle. Santos Dumont sat beside this motor, which worked a propeller, and this curious craft was guided several times by the inventor round the Botanical Gardens in Paris.

About two years after these experiments the science of aeronautics received very valuable aid from M. Deutsch, a member of the French Aero Club. A prize of about L4000 was offered by this gentleman to the man who should first fly from the Aero Club grounds at Longchamps, double round the Eiffel Tower, and then sail back to the starting–place. The total distance to be flown was rather more than 3 miles, and it was stipulated that the journey—which could be made either in a dirigible air–ship or a flying machine—should be completed within half an hour.

This munificent offer at once aroused great enthusiasm among aeronauts and engineers throughout the whole of France, and, to a lesser degree, in Britain. Santos Dumont at once set to work on another air–ship, which was equipped with a much more powerful motor than he had previously used. In July, 1901, his arrangements were completed, and he made his first attempt to win the prize.

The voyage from Longchamps to the Eiffel Tower was made in very quick time, for a favourable wind speeded the huge balloon on its way. The pilot was also able to steer a course round the tower, but his troubles then commenced. The wind was now in his face, and his engine–a small motor engine of about 15 horse–power–was unable to produce sufficient power to move the craft quickly against the wind. The plucky inventor kept fighting against the–breeze, and at length succeeded in returning to his starting–point; but he had exceeded the time limit by several minutes and thus, was disqualified for the prize.

Another attempt was made by Santos Dumont about a month later. This time, however, he was more unfortunate, and he had a marvellous escape from death. As on the previous occasion he got into great difficulties when sailing against the wind on the return journey, and his balloon became torn, so that the gas escaped and the whole craft crashed down on the house–tops. Eyewitnesses of the accident expected to find the gallant young Brazilian crushed to death; but to their great relief he was seen to be hanging to the car, which had been caught upon the buttress of a house. Even now he was in grave peril, but after a long delay he was rescued by means of a rope.

It might be thought that such an accident would have deterred the inventor from making further attempts on the prize; but the aeronaut seemed to be well endowed with the qualities of patience and perseverance and continued to try again. Trial after trial was made, and numerous accidents took place. On nearly every occasion it was comparatively easy to sail round the Tower, but it was a much harder task to sail back again.

At length in October, 1901, he was thought to have completed the course in the allotted time; but the Aero Club held that he had exceeded the time limit by forty seconds. This decision aroused great indignation among Parisians—especially among those who had watched the flight—many of whom were convinced that the journey had been accomplished in the half–hour. After much argument the committee which had charge of the race, acting on the advice of M. Deutsch, who was very anxious that the prize should be awarded to Santos Dumont, decided that the conditions of the flight had been complied with, and that the prize had been legitimately won. It is interesting to read that the famous aeronaut divided the money among the poor.

But important though Santos Dumont's experiments were with the air–ship, they were of even greater value when he turned his attention to the aeroplane.

One of his first trials with a heavier–than–air machine was made with a huge glider, which was fitted with floats. The curious craft was towed along the River Seine by a fast motor boat named the Rapiere, and it actually succeeded in rising into the air and flying behind the boat like a gigantic kite.

12th November, 1906, is a red–letter day in the history of aviation, for it was then that Santos Dumont made his first little flight in an aeroplane. This took place at Bagatelle, not far from Paris.

Two months before this the airman had succeeded in driving his little machine, called the Bird of Prey, many yards into the air, and "11 yards through the air", as the newspapers reported; but the craft was badly smashed. It was not until November that the first really satisfactory flight took place.

A description of this flight appeared in most of the European newspapers, and I give a quotation from one of them: "The aeroplane rose gracefully and gently to a height of about 15 feet above the earth, covering in this most remarkable dash through the air a distance of about 700 feet in twenty–one seconds.

"It thus progressed through the atmosphere at the rate of nearly 30 miles an hour. Nothing like this has ever been accomplished before. . . . The aeroplane has now reached the practical stage."

The dimensions of this aeroplane were:

Length 32 feet Greatest width 39 feet Weight with one passenger 465 pounds. Speed 30 miles an hour

A modern aeroplane with airman and passenger frequently weighs over 1 ton, and reaches a speed of over 60 miles an hour.

It is interesting to note that Santos Dumont, in 1913—that is, only seven years after his flight in an aeroplane at Bagatelle made him world–famous—announced his intention of again taking an active part in aviation. His purpose was to make use of aeroplanes merely for pleasure, much as one might purchase a motor–car for the same object.

Could the intrepid Brazilian in his wildest dreams have foreseen the rapid advance of the last eight years? In 1906 no one had flown in Europe; by 1914 hundreds of machines were in being, in which the pilots were no longer subject to the wind's caprices, but could fly almost where and when they would.

Frenchmen have honoured, and rightly honoured, this gallant and picturesque figure in the annals of aviation, for in 1913 a magnificent monument was unveiled in France to commemorate his pioneer work.

CHAPTER XXVIII. M. Bleriot and the Monoplane

If the Wright brothers can lay claim to the title of "Fathers of the Biplane", then it is certain that M. Bleriot, the gallant French airman, can be styled the "Father of the Monoplane."

For five years—1906 to 1910—Louis Bleriot's name was on everybody's lips in connection with his wonderful records in flying and skilful feats of airmanship. Perhaps the flight which brought him greatest renown was that accomplished in July, 1909, when he was the first man to cross the English Channel by aeroplane. This attempt had been forestalled, although unsuccessfully, by Hubert Latham, a daring aviator who is best known in Lancashire by his flight in 1909 at Blackpool in a wind which blew at the rate of nearly 40 miles an hour—a performance which struck everyone with wonder in these early days of aviation.

Latham attempted, on an Antoinette monoplane, to carry off the prize of L1000 offered by the proprietors of the Daily Mail. On the first occasion he fell in mid–Channel, owing to the failure of his motor, and was rescued by a torpedo–boat. His machine was so badly damaged during the salving operations that another had to be sent from Paris, and with this he made a second attempt, which was also unsuccessful. Meanwhile M. Bleriot had arrived on the scene; and on 25th July he crossed the Channel from Calais to Dover in thirty–seven minutes and was awarded the L1000 prize.

Bleriot's fame was now firmly established, and on his return to France he received a magnificent welcome. The monoplane at once leaped into favour, and the famous "bird man" had henceforth to confine his efforts to the building of machines and the organization of flying events. He has since established a large factory in France and inaugurated a flying school at Pau.

All the time that the Wrights were experimenting with their glider and biplane in America, and the Voisin brothers were constructing biplanes in France, Bleriot had been giving earnest attention to the production of a real "bird" machine, provided with one pair of FLAPPING wings. We know now that such an aeroplane is not likely to be of practical use, but with quiet persistence Bleriot kept to his task, and succeeded in evolving the famous Antoinette monoplane, which more closely resembles a bird than does any other form of air–craft.

In the illustration of the Bleriot monoplane here given you will notice that there is one main plane, consisting of a pair of highly–cambered wings; hence the name "MONOplane". At the rear of the machine there is a much smaller plane, which is slightly cambered; this is the elevating plane, and it can be tilted up or down in order to raise or lower the machine. Remember that the elevating plane of a biplane is to the front of the machine and in the monoplane at the rear. The small, upright plane G is the rudder, and is used for steering the machine to the right or left. The long narrow body or framework of the monoplaneis known as the FUSELAGE.

By a close study of the illustration, and the description which accompanies it, you will understand how the machine is driven. The main plane is twisted, or warped, when banking, much in the same way that the Wright biplane is warped.

Far greater speed can be obtained from the monoplane than from the biplane, chiefly because in the former machine there is much less resistance to the air. Both height and speed records stand to the credit of the monoplane.

The enormous difference in the speeds of monoplanes and biplanes can be best seen at a race meeting at some aerodrome. Thus at Hendon, when a speed handicap is in progress, the slow biplanes have a start of one or two laps over the rapid little monoplanes in a six–lap contest, and it is most amusing to see the latter dart under, or over, the more cumbersome biplane. Recently however, much faster biplanes have been built, and they bid fair to rival the swiftest monoplanes in speed.

There is, however, one serious drawback to the use of the monoplane: it is far more dangerous to the pilot than is the biplane. Most of the fatal accidents in aviation have been caused through mishaps to monoplanes or their engines, and chiefly for this reason the biplane has to a large extent supplanted the monoplane in warfare. The biplane, too, is better adapted for observation work, which is, after all, the chief use of air–craft.

In a later chapter some account will be givcn of the three types of aeroplane which the war has evolved—the general–purposes machine, the single–seater "fighter", and those big bomb–droppers, the British Handley Page and the German Gotha.

CHAPTER XXIX. Henri Farman and the Voisin Biplane

The coming of the motor engine made events move rapidly in the world of aviation. About the year 1906 people's attention was drawn to France, where Santos Dumont was carrying out the wonderful experiments which we have already described. Then came Henri Farman, who piloted the famous biplane built by the Voisin brothers in 1907; an aeroplane destined to bring world–wide renown to its clever constructors and its equally clever and daring pilot.

There were notable points of distinction between the Voisin biplane and that built by the Wrights. The latter, as we have seen, had two propellers; the former only one. The launching skids of the Wright biplane gave place to wheels on Farman's machine. One great advantage, however, possessed by the early Wright biplane over its French rivals, was in its greater general efficiency. The power of the engine was only about one–half of the power required in certain of the French designs. This was chiefly due to the use of the launching rail, for it needed much greater motor power to make a machine rise from the ground by its own motor engine than when it received a starting lift from a falling weight. Even in our modern aeroplanes less engine power is required to drive the craft through the air than to start from the ground.

Farman achieved great fame through his early flights, and, on 13th January, 1908, at the flying ground at Issy, in France, he won the prize of L2000, offered by MM. Deutsch and Archdeacon to the first aviator who flew a circular kilometre. In July of the same year he won another substantial prize given by a French engineer, M. Armengaud, to the first pilot who remained aloft for a quarter of an hour.

Probably an even greater performance was the cross–country flight made by Farman about three months later. In the flight he passed over hills, valleys, rivers, villages, and woods on his journey from Chalons to Rheims, which he accomplished in twenty minutes.

In the early models of the Voisin machine there were fitted between the two main planes a number of vertical planes, as shown clearly in the illustration facing p. 160. It was thought that these planes would increase the stability of the machine, independent of the skill of the operator, and in calm weather they were highly effective. Their great drawback, however, was that when a strong side wind caught them the machine was blown out of its course.

Subsequently Farman considerably modified the early–type Voisin biplane, as shown by the illustration facing p. 160. The vertical planes were dispensed with, and thus the idea of automatic stability was abandoned.

But an even greater distinction between the Farman biplane and that designed by the Wrights was in the adoption of a system of small movable planes, called AILERONS, fixed at extremities of the main planes, instead of the warping controls which we have already described. The ailerons, which are adapted to many of our modern aeroplanes, are really balancing flaps, actuated by a control lever at the right side of the pilot's seat, and the principle on which they are worked is very similar to that employed in the warp system of lateral stability.

CHAPTER XXX. A Famous British Inventor

About the time that M. Bleriot was developing his monoplane, and Santos Dumont was astonishing the world with his flying feats at Bagatelle, a young army officer was at work far away in a secluded part of the Scottish Highlands on the model of an aeroplane. This young man was Lieutenant J. W. Dunne, and his name has since been on everyone's lips wherever aviation is discussed. Much of Lieutenant Dunne's early experimental work was done on the Duke of Atholl's estate, and the story goes that such great secrecy was observed that "the tenants were enrolled as a sort of bodyguard to prevent unauthorized persons from entering". For some time the War Office helped the inventor with money, for the numerous tests and trials necessary in almost every invention before satisfactory results are achieved are very costly.

Probably the inventor did not make sufficiently rapid progress with his novel craft, for he lost the financial help and goodwill of the Government for a time; but he plodded on, and at length his plans were sufficiently advanced for him to carry on his work openly. It must be borne in mind that at the time Dunne first took up the study of aviation no one had flown in Europe, and he could therefore receive but little help from the results achieved by other pilots and constructors.

But in the autumn of 1913 Lieutenant Dunne's novel aeroplane was the talk of both Europe and America. Innumerable trials had been made in the remote flying ground at Eastchurch, Isle of Sheppey, and the machine became so far advanced that it made a cross–Channel flight from Eastchurch to Paris. It remained in France for some time, and Commander Felix, of the French Army, made many excellent flights in it. Unfortunately, however, when flying near Deauville, engine trouble compelled the officer to descend; but in making a landing in a very small field, not much larger than a tennis–court, several struts of the machine were damaged. It was at once seen that the aeroplane could not possibly be flown until it had been repaired and thoroughly overhauled. To do this would take several days, especially as there were no facilities for repairing the craft near by, and to prevent anyone from making a careful examination of the aeroplane, and so discovering the secret features which had been so jealously guarded, the machine was smashed up after the engine had been removed.

At that time this was the only Dunne aeroplane in existence, but of course the plans were in the possession of the inventor, and it was an easy task to make a second machine from the same model. Two more machines were put in hand at Hendon, and a third at Eastchurch.

On 18th October, 1913, the Dunne aeroplane made its first public appearance at Hendon, in the London aerodrome, piloted by Commander Felix. The most striking distinction between this and other biplanes is that its wings or planes, instead of reaching from side to side of the engine, stretch back in the form of the letter V, with the point of the V to the front. These wings extend so far to the rear that there is no need of a tail to the machine, and the elevating plane in front can also be dispensed with.

This curious and unique design in aeroplane construction was decided upon by Lieutenant Dunne after a prolonged observation at close quarters of different birds in flight, and the inventor claims for his aeroplane that it is practically uncapsizable. Perhaps, however, this is too much to claim for any heavier–than–air machine; but at all events the new design certainly appears to give greater stability, and it is to be hoped that by this and other devices the progress of aviation will not in the future be so deeply tinged with tragedy.

CHAPTER XXXI. The Romance of a Cowboy Aeronaut

In the brief but glorious history of pioneer work in aviation, so far as it applies to this country, there is scarcely a more romantic figure to be found than Colonel Cody. It was the writer's pleasure to come into close contact with Cody during the early years of his experimental work with man–lifting box–kites at the Alexandra Park, London, and never will his genial smile and twinkling eye be forgotten.

Cody always seemed ready to crack a joke with anyone, and possibly there was no more optimistic man in the whole of Britain. To the boys and girls of Wood Green he was a popular hero. He was usually clad in a "cowboy" hat, red flannel shirt, and buckskin breeches, and his hair hung down to his shoulders. On certain occasions he would give a "Wild West" exhibition at the Alexandra Palace, and one of his most daring tricks with the gun was to shoot a cigarette from a lady's lips. One could see that he was entire master of the rifle, and a trick which always brought rounds of applause was the hitting of a target while standing with his back to it, simply by the aid of a mirror held at the butt of his rifle.

But it is of Cody as an aviator and aeroplane constructor that we wish to speak. For some reason or other he was generally the object of ridicule, both in the Press and among the public. Why this should have been so is not quite clear; possibly his quaint attire had something to do with it, and unfriendly critics frequently raised a laugh at his expense over the enormous size of his machines. So large were they that the Cody biplane was laughingly called the "Cody bus" or the "Cody Cathedral."

But in the end Cody fought down ridicule and won fame, for in competition with some of the finest machines of the day, piloted by some of our most expert airmen, he won the prize of L5000 offered by the Government in 1912 in connection with the Army trials for aeroplanes. In these trials he astonished everyone by obtaining a speed of over 70 miles an hour in his biplane, which weighed 2600 pounds.

In the opening years of the present century Cody spent much time in demonstrations with huge box–kites, and for a time this form of kite was highly popular with boys of North London. In these kites he made over two hundred flights, reaching, on some occasions, an altitude of over 2000 feet. At all times of the day he could have been seen on the slopes of the Palace Hill, hauling these strange–looking, bat–like objects backward and forward in the wind. Reports of his experiments appeared in the Press, but Cody was generally looked upon as a "crank". The War Office, however, saw great possibilities in the kites for scouting purposes in time of war, and they paid Cody L5000 for his invention.

It is a rather romantic story of how Cody came to take up experimental work with kites, and it is repeated as it was given by a Mohawk chief to a newspaper representative.

"On one occasion when Cody was in a Lancashire town with his Wild West show, his son Leon went into the street with a parrot–shaped kite. Leon was attired in a red shirt, cowboy trousers, and sombrero, and soon a crowd of youngsters in clogs was clattering after him.

"'If a boy can interest a crowd with a little kite, why can't a man interest a whole nation?' thought Cody—and so the idea of man–lifting kites developed."

In 1903 Cody made a daring but unsuccessful attempt to cross the Channel in a boat drawn by two kites. Had he succeeded he intended to cross the Atlantic by similar means.

Later on, Cody turned his attention to the construction of aeroplanes, but he was seriously handicapped by lack of funds. His machines were built with the most primitive tools, and some of our modern constructors, working in well-equipped "shops", where the machinery is run by electric plant, would marvel at the work accomplished with such tools as those used by Cody.

Most of Cody's flights were made on Laffan's Plain, and he took part in the great "Round Britain" race in 1911. It was characteristic of the man that in this race he kept on far in the wake of MM. Beaumont and Vedrines, though he knew that he had not the slightest chance of winning the prize; and, days after the successful pilot had arrived back at Brooklands, Cody's "bus" came to earth in the aerodrome. "It's dogged as does it," he remarked, "and I meant to do the course, even if I took a year over it."

Of Cody's sad death at Farnborough, when practising in the ill-fated water-plane which he intended to pilot in the sea flight round Great Britain in 1913, we speak in a later chapter.

CHAPTER XXXII. Three Historic Flights

When the complete history of aviation comes to be written, there will be three epoch-making events which will doubtless be duly appreciated by the historian, and which may well be described as landmarks in the history of flight. These are the three great contests organized by the proprietors of the Daily Mail, respectively known as the "London to Manchester" flight, the "Round Britain flight in an aeroplane", and the "Water-plane flight round Great Britain."

In any account of aviation which deals with the real achievements of pioneers who have helped to make the science of flight what it is to-day, it would be unfair not to mention the generosity of Lord Northcliffe and his co-directors of the Daily Mail towards the development of aviation in this country. Up to the time of writing, the sum of L24,750 has been paid by the Daily Mail in the encouragement of flying, and prizes to the amount of L15,000 are still on offer. In addition to these prizes this journal has maintained pilots who may be described as "Missionaries of Aviation". Perhaps the foremost of them is M. Salmet, who has made hundreds of flights in various parts of the country, and has aroused the greatest enthusiasm wherever he has flown.

The progress of aviation undoubtedly owes a great deal to the Press, for the newspaper has succeeded in bringing home to most people the fact that the possession of air–craft is a matter of national importance. It was of little use for airmen to make thrilling flights up and down an aerodrome, with the object of interesting the general public, if the newspapers did not record such flights, and though in the very early days of aviation some newspapers adopted an unfriendly attitude towards the possibilities of practical aviation, nearly all the Press has since come to recognize the aeroplane as a valuable means of national defence. Right from the start the Daily Mail foresaw the importance of promoting the new science of flight by the award of prizes, and its public–spirited enterprise has done much to break up the prevailing apathy towards aviation among the British nation.

If these three great events had been mere spectacles and nothing else—such as, for instance, that great horse–race known as "The Derby"—this chapter would never have been written. But they are most worthy of record because all three have marked clearly–defined stepping–stones in the progress of flight; they have proved conclusively that aviation is practicable, and that its ultimate entry into the busy life of the world is no more than a matter of perfecting details.

The first L10,000 prize was offered in November, 1906, for a flight by aeroplane from London to Manchester in twenty–four hours, with not more than two stoppages en route. In 1910 two competitors entered the lists for the flight; one, an Englishman, Mr. Claude Grahame–White; the other, a Frenchman, M. Paulhan.

Mr. Grahame–White made the first attempt, and he flew remarkably well too, but he was forced to descend at Lichfield—about 113 miles on the journey—owing to the high and gusty winds which prevailed in the Trent valley. The plucky pilot intended to continue the flight early the next morning, but during the night his biplane was blown over in a gale while it stood in a field, and it was so badly damaged that the machine had to be sent back to London to be repaired.

This took so long that his French rival, M. Paulhan, was able to complete his plans and start from Hendon, on 27th April. So rapidly had Paulhan's machine been transported from Dover, and "assembled" at Hendon, that Mr. White, whose biplane was standing ready at Wormwood Scrubbs, was taken by surprise when he heard that his rival had started on the journey and "stolen a march on him", so to speak. Nothing daunted, however, the plucky British aviator had his machine brought out, and he went in pursuit of Paulhan late in the afternoon. When darkness set in Mr. White had reached Roade, but the French pilot was several miles ahead.

Now came one of the most thrilling feats in the history of aviation. Mr. White knew that his only chance of catching Paulhan was to make a flight in the darkness, and though this was extremely hazardous he arose from a small field in the early morning, some hours before daybreak arrived, and flew to the north. His friends had planned ingenious devices to guide him on his way: thus it was proposed to send fast motor–cars, bearing very powerful lights, along the route, and huge flares were lighted on the railway; but the airman kept to his course chiefly by the help of the lights from the railway stations.

Over hill and valley, forest and meadow, sleeping town and slumbering village, the airman flew, and when dawn arrived he had nearly overhauled his rival, who, in complete ignorance of Mr. White's daring pursuit, had not yet started.

But now came another piece of very bad luck for the British aviator. At daybreak a strong wind arose, and Mr. White's machine was tossed about like a mere play–ball, so that he was compelled to land. Paulhan, however, who was a pilot with far more experience, was able to overcome the treacherous air gusts, and he flew on to Manchester, arriving there in the early morning.

Undoubtedly the better pilot won, and he had a truly magnificent reception in Manchester and London, and on his return to France. But this historic contest laid the foundation of Mr. Grahame–White's great reputation as an aviator, and, as we all know, his fame has since become world–wide.

CHAPTER XXXIII. Three Historic Flights (Cont.)

About a month after Paulhan had won the "London to Manchester" race, the world of aviation, and most of the general public too, were astonished to read the announcement of another enormous prize. This time a much harder task was set, for the conditions of the contest stated that a circuit of Britain had to be made, covering a distance of about 1000 miles in one week, with eleven compulsory stops at fixed controls.

This prize was offered on 22nd May, 1910, and in the following year seventeen competitors entered the lists. It says much for the progress of aviation at this time, when we read that, only a year before, it was difficult to find but two pilots to compete in the much easier race described in the last chapter. Much of this progress was undoubtedly due to the immense enthusiasm aroused by the success of Paulhan in the "London to Manchester" race.

We will not describe fully the second race, because, though it was of immense importance at the time, it has long since become a mere episode. Rarely has Britain been in such great excitement as during that week in July, 1911.

Engine troubles, breakdowns, and other causes soon reduced the seventeen competitors to two only: Lieutenant Conneau, of the French Navy–who flew under the name of M. Beaumont—and M. Vedrines. Neck to neck they flew—if we may be allowed this horse–racing expression—over all sorts of country, which was quite unknown to them.

Victory ultimately rested with Lieutenant Conneau, who, on 26th July, 1911, passed the winning–post at Brooklands after having completed the course in the magnificent time of twenty–two hours, twenty–eight minutes, averaging about 45 miles an hour for the whole journey. M. Vedrines, though defeated, made a most plucky fight. Conneau's success was due largely to his ability to keep to the course—on two or three occasions Vedrines lost his way— and doubtless his naval training in map–reading and observation gave him the advantage over his rival.

The third historic flight was made by Mr. Harry Hawker, in August, 1913. This was an attempt to win a prize of L5000 offered by the proprietors of the Daily Mail for a flight round the British coasts. The route was from Cowes, in the Isle of Wight, along the southern and eastern coasts to Aberdeen and Cromarty, thence through the Caledonian Canal to Oban, then on to Dublin, thence to Falmouth, and along the south coast to Southampton Water.

Two important conditions of the contest were that the flight was to be made in an all–British aeroplane, fitted with a British engine. Hitherto our aeroplane constructors and engine companies were behind their rivals across the Channel in the building of air–craft and aerial engines, and this country freely acknowledged the merits and enterprise of French aviators. Though in the European War it was afterwards proved that the British airman and constructor were the equals if not the superiors of any in the world, at the date of this contest they were behind in many respects.

As these conditions precluded the use of the famous Gnome engine, which had won so many contests, and indeed the employment of any engine made abroad, the competitors were reduced to two aviation firms; and as one or these ultimately withdrew from the contest the Sopwith Aviation Company of Kingston–on–Thames and Brooklands entered a machine.

Mr. T. Sopwith chose for his pilot a young Australian airman, Mr. Harry Hawker. This skilful airman came with three other Australians to this country to seek his fortune about three years before. He was passionately devoted to mechanics, and, though he had had no opportunity of flying in his native country, he had been intensely interested in the progress of aviation in France and Britain, and the four friends set out on their long journey to seek work in aeroplane factories.

All four succeeded, but by far the most successful was Harry Hawker. Early in 1913 Mr. Sopwith was looking out for a pilot, and he engaged Hawker, whom he had seen during some good flying at Brooklands.

In a month or two he was engaged in record breaking, and in June, 1913, he tried to set up a new British height record. In his first attempt he rose to 11,300 feet; but as the carburettor of the engine froze, and as the pilot himself was in grave danger of frost–bite, he descended. About a fortnight later he rose 12,300 feet above sea–level, and shortly afterwards he performed an even more difficult test, by climbing with three passengers to an altitude of 8500 feet.

With such achievements to his name it was not in the least surprising that Mr. Sopwith's choice of a pilot for the water–plane race rested on Hawker. His first attempt was made on 16th August, when he flew from Southampton Water to Yarmouth—a distance of about 240 miles—in 240 minutes. The writer, who was spending a holiday at Lowestoft, watched Mr. Hawker go by, and his machine was plainly visible to an enormous crowd which had lined the beach.

To everyone's regret the pilot was affected with a slight sunstroke when he reached Yarmouth, and another Australian airman, Mr. Sidney Pickles, was summoned to take his place. This was quite within the rules of the contest, the object of which was to test the merits of a British machine and engine rather than the endurance and skill of a particular pilot. During the night a strong wind arose, and next morning, when Mr. Pickles attempted to resume the flight, the sea was too rough for a start to be made, and the water–plane was beached at Gorleston.

Mr. Hawker quickly recovered from his indisposition, and on Monday, 25th August, he, with a mechanic as passenger, left Cowes about five o'clock in the morning in his second attempt to make a circuit of Britain. The first control was at Ramsgate, and here he had to descend in order to fulfil the conditions of the contest.

Ramsgate was left at 9.8, and Yarmouth, the next control, was reached at 10.38. So far the engine, built by Mr. Green, had worked perfectly. About an hour was spent at Yarmouth, and then the machine was en route to Scarborough. Haze compelled the pilot to keep close in to the coast, so that he should not miss the way, and a choppy breeze some what retarded the progress of the machine along the east coast. About 2.40 the pilot brought his machine to earth, or rather to water, at Scarborough, where he stayed for nearly two hours.

Mr. Hawker's intention was to reach Aberdeen, if possible, before nightfall, but at Seaham he had to descend for water, as the engine was becoming uncomfortably hot, and the radiator supply of water was rapidly diminishing. This lost much valuable time, as over an hour was spent here, and it had begun to grow dark before the journey was recommenced. About an hour after resuming his journey he decided to plane down at the fishing village of Beadwell, some 20 miles south of Berwick.

At 8.5 on Tuesday morning the pilot was on his way to Aberdeen, but he had to descend and stay at Montrose for about half an hour, and Aberdeen was reached about 11 a.m. His Scottish admirers, consisting of quite 40,000 people at Aberdeen alone, gave him a most hearty welcome, and sped him on his way about noon. Some two hours later Cromarty was reached.

Now commenced the most difficult part of the course. The Caledonian Canal runs among lofty mountains, and the numerous air–eddies and swift air–streams rushing through the mountain passes tossed the frail craft to and fro, and at times threatened to wreck it altogether. On some occasions the aeroplane was tossed up over 1000 feet at one blow; at other times it was driven sideways almost on to the hills. From Cromarty to Oban the journey was only about 96 miles, but it took nearly three hours to fly between these places. This slow progress seriously jeopardized the pilot's chances of completing the course in the allotted time, for it was his intention to make the coast of Ireland by nightfall. But as it was late when Oban was reached he decided to spend the night there.

Early the following morning he left for Dublin, 222 miles away. Soon a float was found to be waterlogged and much valuable time was, spent in bailing it dry. Then a descent had to be made at Kiells, in Argyllshire, because a valve had gone wrong. Another landing was made at Larne, to take aboard petrol. As soon as the petrol tanks were filled and the machine had been overhauled the pilot got on his way for Dublin.

For over two hours he flew steadily down the Irish coast, and then occurred one of those slight accidents, quite insignificant in themselves, but terribly disastrous in their results. Mr. Hawker's boots were rubber soled and his foot slipped off the rudder bar, so that the machine got out of control and fell into the sea at Lough Shinny, about 15 miles north of Dublin. At the time of the accident the pilot was about 50 feet above the water, which in this part of the Lough is very shallow. The machine was completely wrecked, and Mr. Hawker's mechanic was badly cut about the head and neck, besides having his arm broken. Mr. Hawker himself escaped injury.

All Britons deeply sympathized with his misfortune, and much enthusiasm, was aroused when the proprietors of the Daily Mail presented the skilful and courageous pilot with a cheque for L1000 as a consolation gift.

In a later chapter some account will be given of the tremendous development of the aeroplane during four years of war. But it is fitting that to the three historic flights detailed above there should be added the sensational exploits of the Marchese Giulio Laureati in 1917. This intrepid Italian airman made a non–stop journey from Turin to Naples and back, a distance of 920 miles. A month later he flew from Turin to Hounslow, a distance of 656 miles, in 7 hours 22 minutes. His machine was presented to the British Air Board by the Italian Government.

CHAPTER XXXIV. The Hydroplane and Air–boat

One of the most recent developments in aviation is the hydroplane, or water–plane as it is most commonly called. A hydroplane is an aeroplane fitted with floats instead of wheels, so that it will rise from, or alight upon, the surface of the water. Often water–planes have their floats removed and wheels affixed to the chassis, so that they may be used over land.

From this you may think that the construction of a water–plane is quite a simple task; but such is not the case. The fitting of floats to an aeroplane has called for great skill on the part of the constructor, and many difficulties have had to be overcome.

Those of you who have seen an acroplane rise from the ground know that the machine runs very quickly over the earth at a rapidly– increasing speed, until sufficient momentum is obtained for the machine to lift itself into the air. In the case of the water–plane the pilot has to glide or "taxi" by means of a float or floats over the waves until the machine acquires flying speed.

Now the land resistance to the rubber–tired wheels is very small when compared with the water resistance to the floats, and the faster the craft goes the greater is the resistance. The great problem which the constructor has had to solve is to build a machine fitted with floats which will leave the water easily, which will preserve the lateral balance of the machine, and which will offer the minimum resistance in the air.

A short flat–bottomed float, such as that known as the Fabre, is good at getting off from smooth water, but is frequently damaged when the sea is rough. A long and narrow float is preferable for rough water, as it is able to cut through the waves; but comparatively little "lift" is obtained from it.

Some designers have provided their water–planes with two floats; others advocate a single loat. The former makes the machine more stable when at rest on the water, but a great rawback is that the two–float machine is affected by waves more than a machine fitted with a single float; for one float may be on the crest of a wave and the other in the dip. This is not the case with the single–float water–plane, but on the other hand this type is less stable than the other when at rest.

Sometimes the floats become waterlogged, and so add considerably to the weight of the machine. Thus in Mr. Hawker's flight round Britain, the pilot and his passenger had to pump about ten gallons of water out of one of the floats before the machine could rise properly. Floats are usually made with watertight compartments, and are composed of several thin layers of wood, riveted to a wooden framework.

There is another technical question to be considered in the fixing of the floats, namely, the fore–and–aft balance of the machine in the air. The propeller of a water–plane has to be set higher than that of a land aeroplane, so that it may not come into contact with the waves. This tends to tip the craft forwards, and thus make the nose of the float dig in the water. To overcome this the float is set well forward of the centre of gravity, and though this counteracts the thrust when the craft "taxies" along the waves, it endangers its fore–and–aft stability when aloft.

CHAPTER XXXV. A Famous British Inventor of the Water–plane

Though Harry Hawker made such a brilliant and gallant attempt to win the L5000 prize, we must not forget that great credit is due to Mr. Sopwith, who designed the water–plane, and to Mr. Green, the inventor of the engine which made such a flight possible, and enabled the pilot to achieve a feat never before approached in any part of the world.

The life–story of Mr. "Tommy" Sopwith is almost a romance. As a lad he was intensely interested in mechanics, and we can imagine him constructing all manner of models, and enquiring the why and the wherefore of every mechanical toy with which he came into contact.

At the early age of twenty–one he commenced a motor business, but about this time engineers and mechanics all over the country were becoming greatly interested in the practical possibilities of aviation. Mr. Sopwith decided to learn to fly, and in 1910, after continued practice in a Howard Wright biplane, he had become a proficient pilot. So rapid was his progress that by the end of the year he had won the magnificent prize of L4000 generously offered by Baron de Forest for the longest flight made by an all–British machine from England to the Continent. In this flight he covered 177 miles, from Eastchurch, Isle of Sheppey, to the Belgian frontier, in three and a half hours.

If Mr. Sopwith had been in any doubt as to the wisdom of changing his business this remarkable achievement alone must have assured him that his future career lay in aviation. In 1911 he was graciously received by King George V at Windsor Castle, after having flown from Brooklands and alighted on the East Terrace of the famous castle.

In the same year he visited America, and astonished even that go–ahead country with some skilful flying feats. To show the practical possibilities of the aeroplane he overtook the liner Olympic, after she had left New York harbour on her homeward voyage, and dropped aboard a parcel addressed to a passenger. On his return to England he competed in the first Aerial Derby, the course being a circuit of London, representing a distance of 81 miles. In this race he made a magnificent flight in a 70–horse–power Bleriot monoplane, and came in some fifteen minutes before Mr. Hamel, the second pilot home. So popular was his victory that Mr. Grahame–White and several other officials of the London Aerodrome carried him shoulder high from his machine.

From this time we hear little of Mr. Sopwith as a pilot, for, like other famous airmen, such as Louis Bleriot, Henri Farman, and Claude Grahame–White, who jumped into fame by success in competition flying, he has retired with his laurels, and now devotes his efforts to the construction of machines. He bids fair to be equally successful as a constructor of air–craft as he formerly was as a pilot of flying machines. The Sopwith machines are noted for their careful design and excellent workmanship. They are made by the Sopwith Aviation Company, Ltd., whose works are at Kingston–on–Thames. Several water–planes have been built there for the Admiralty, and land machines for the War Office. Late in 1913 Mr. Hawker left Britain for Australia to give demonstrations in the Sopwith machine to the Government of his native country.

A fine list of records has for long stood to the credit of the Sopwith biplane. Among these are:

```
British Height Record (Pilot only) ...     ...  11,450 feet
   "       "      "   (Pilot and 1 Passenger)  12,900  "
   "       "      "   (Pilot and 2 Passengers) 10,600  "
World's    "      "   (Pilot and 3 Passengers)  8,400  "
```

Many of the Sopwith machines used in the European War were built specially to withstand rough climate and heavy winds, and thus they were able to work in almost every kind of weather. It was this fact, coupled with the indomitable spirit of adventure inherent in men of British race, that made British airmen more than hold their own with both friend and foe in the war.

CHAPTER XXXVI. Sea–planes for Warfare

"Even in the region of the air, into which with characteristic British prudence we have moved with some tardiness, the Navy need not fear comparison with the Navy of any other country. The British sea–plane, although still in an empirical stage, like everything else in this sphere of warlike operations, has reached a point of progress in advance of anything attained elsewhere.

"Our hearts should go out to–night to those brilliant officers, Commander Samson and his band of brilliant pioneers, to whose endeavours, to whose enterprise, to whose devotion it is due that in an incredibly short space of time our naval aeroplane service has been raised to that primacy from which it must never be cast down.

"It is not only in naval hydroplanes that we must have superiority. The enduring safety of this country will not be maintained by force of arms unless over the whole sphere of aerial development we are able to make ourselves the first nation. That will be a task of long duration. Many difficulties have to be overcome. Other countries have started sooner. The native genius of France, the indomitable perseverance of Germany, have produced results which we at the present time cannot equal."

So said Mr. Winston Churchill at the Lord Mayor's Banquet held in London in 1913, and I have quoted his speech because such a statement, made at such a time, clearly shows the attitude of the British Government toward this new arm of Imperial Defence.

In bygone days the ocean was the great highway which united the various quarters of the Empire, and, what was even more important from the standpoint of our country's defence, it was a formidable barrier between Britain and her Continental neighbours,

```
"Which serves it in the office of a wall
 Or as a moat defensive to a house."
```

But the ocean is no longer the only highway, for the age of aerial navigation has arrived, and, as one writer says: "Every argument which impelled us of old to fight for the dominion of the sea has apparently been found valid in relation to the supremacy of the air."

From some points of view this race between nations for naval and aerial supremacy may be unfortunate, but so long as the fighting instinct of man continues in the human race, so long as rivalry exists between nations, so long must we continue to strengthen our aerial position.

Britain is slow to start on any great venture where great change is effected. Our practice is rather to wait and see what other nations are doing; and there is something to be said for this method of procedure.

In the art of aviation, and in the construction of air–craft, our French, German, and American rivals were very efficient pacemakers in the aerial race for supremacy, and during the years 1909–12 we were in grave peril of being left hopelessly behind. But in 1913 we realized the vital importance to the State of capturing the first place in aviation, particularly that of aerial supremacy at sea, for the Navy is our first line of defence. So rapid has been our progress that we are quite the equal of our French and German rivals in the production of aeroplanes, and in sea–planes we are far ahead of them, both in design and construction, and the war has proved that we are ahead in the art of flight.

The Naval Air Service before the war had been establishing a chain of air stations round the coast. These stations are at Calshot, on Southampton Water, the Isle of Grain, off Sheerness, Leven, on the Firth of Forth, Cromarty, Yarmouth, Blythe, and Cleethorpes.

But what is even more important is the fact that the Government is encouraging sea–plane constructors to go ahead as fast as they can in the production of efficient machines. Messrs. Short Brothers, the Sopwith Aviation Company, and Messrs. Roe are building high–class machines for sea work which can beat anything turned out abroad. Our newest naval water–planes are fitted with British–built wireless apparatus of great range of action, and Messrs. Short Brothers are at the present time constructing for the Admiralty, at their works in the Isle of Sheppey, a fleet of fighting water–planes capable of engaging and destroying the biggest dirigible air–ships.

In 1913 aeroplanes took a very prominent part in our naval manoeuvres, and the cry of the battleship captains was: "Give us water–planes. Give us them of great size and power, large enough to carry a gun and gun crew, and capable of taking twelve–hour cruises at a speed much greater than that of the fastest dirigible air–ship, and we shall be on the highroad to aerial supremacy at sea."

The Admiralty, acting on this advice, at once began to co–operate with the leading firms of aeroplane constructors, and at a great rate machines of all sizes and designs have been turned out. There were light single–seater water–planes able to maintain a speed of over a mile a minute; there were also larger machines for long–distance flying which could carry two passengers. The machines were so designed that their wings could be folded back along their bodies, and their wires, struts, and so on packed into the main parts of the craft, so that they were almost as compact as the body of a bird at rest on its perch, and they took up comparatively little space on board ship.

A brilliantly executed raid was carried out on Cuxhaven, an important German naval base, by seven British water–planes, on Christmas Day, 1914. The water–planes were escorted across the North Sea by a light cruiser and destroyer force, together with submarines. They left the war–ships in the vicinity of Heligoland and flew over Cuxhaven, discharging bombs on points of military significance, and apparently doing considerable damage to the docks and shipping. The British ships remained off the coast for three hours in order to pick up the returning airmen, and during this time they were attacked by dirigibles and submarines, without, however, suffering damage. Six of the sea–planes returned safely to the ships, but one was wrecked in Heligoland Bight.

But the present efficient sea–plane is a development of the war. In the early days many of the raids of the "naval wing" were carried out in land–going aeroplanes. Now the R.N.A.S., which came into being as a separate service in July, 1914, possess two main types of flying machine, the flying boat and the twin float, both types being able to rise from and alight upon the sea, just as an aeroplane can leave and return to the land. Many brilliant raids stand to the credit of the R.N.A.S. The docks at Antwerp, submarine bases at Ostend, and all Germany's fortified posts on the Belgian coast, have seldom been free from their attentions. And when, under the stress of public outcry, the Government at last gave its consent to a measure of "reprisals" it was the R.N.A.S. which opened the campaign with a raid upon the German town of Mannheim.

As the war continued the duties of the naval pilot increased. He played a great part in the ceaseless hunt for submarines. You must often have noticed how easily fish can be seen from a bridge which are quite invisible from the banks of the river. On this principle the submarine can be "spotted" by air–craft, and not until the long silence upon naval affairs is broken, at the end of the war, shall we know to what extent we are indebted to naval airmen for that long list of submarines which, in the words of the German reports, "failed to return" to their bases.

In addition to the "Blimps" of which mention has been made, the Royal Naval Air Service are in charge of air–ships known as the Coast Patrol type, which work farther out to sea, locating minefields and acting as scouts for the great fleet of patrol vessels. The Service has gathered laurels in all parts of the globe, its achievements ranging from an aerial food service into beleaguered Kut to the discovery of the German cruiser Konigsberg, cunningly camouflaged up an African creek.

CHAPTER XXXVII. The First Man to Fly in Britain

The honour of being the first man to fly in this country is claimed by Mr. A. V. Roe, head of the well–known firm A. V. Roe Co., of Manchester, and constructor of the highly–efficient Avro machines.

As a youth Roe's great hobby was the construction of toy models of various forms of machinery, and later on he achieved considerable success in the production of aeroplane models. All manner of novelties were the outcome of his fertile brain, and as it has been truly remarked, "his novelties have the peculiarity, not granted to most pioneers, of being in one respect or another ahead of his contemporaries." In addition, he studied the flight of birds.

In the early days of aviation Mr. Roe was a firm believer in the triplane form of machine, and his first experiments in flight were made with a triplane equipped with an engine which developed only 9 horse–power.

Later on, he turned his attention to the biplane, and with this craft he has been highly successful. The Avro biplane, produced in 1913, was one of the very best machines which appeared in that eventful year. The Daily Telegraph, when relating its performances, said: "The spectators at Hendon were given a remarkable demonstration of the wonderful qualities of this fine Avro biplane, whose splendid performances stamped it as one of the finest aeroplanes ever designed, if not indeed the finest of all".

This craft is fitted with an 80–horse–power Gnome engine, and is probably the fastest passenger–carrying biplane of its type in the world. Its total weight, with engine, fuel for three hours, and a passenger, is 1550 pounds, and it has a main–plane surface of 342 square feet.

Not only can the biplane maintain such great speed, but, what is of great importance for observation purposes, it can fly at the slow rate of 30 miles per hour. We have previously remarked that a machine is kept up in the air by the speed it attains; if its normal flying speed be much reduced the machine drops to earth unless the rate of flying is accelerated by diving, or other means.

What Harry Hawker is to Mr. Sopwith so is F. P. Raynham to Mr. Roe. This skilful pilot learned to fly at Brooklands, and during the last year or two he has been continuously engaged in testing Avro machines, and passing them through the Army reception trials. In the "Aerial Derby" of 1913 Mr. Raynham piloted an 80–horse–power Avro biplane, and came in fourth.

CHAPTER XXXVIII. The Royal Flying Corps and Royal Naval Air Service

The year 1912 was marked by the institution of the Royal Flying Corps. The new corps, which was so soon to make its mark in the greatest of all wars, consisted of naval and military "wings". In those early days the head–quarters of the corps were at Eastchurch, and there both naval and military officers were trained in aviation. In an arm of such rapid—almost miraculous—development as Service flying to go back a period of six years is almost to take a plunge into ancient history. Designs, engines, guns, fittings, signals of those days are now almost archaic. The British engine of reliable make had not yet been evolved, and the aeroplane generally was a conglomerate affair made up of parts assembled from various parts of the Continent. The present–day sea–plane was yet to come, and naval pilots shared the land–going aeroplanes of their military brethren. In the days when Bleriot provided a world sensation by flying across the Channel the new science was kept alive mainly by the private enterprise of newspapers and aeroplane manufacturers. The official attitude, as is so often the case in the history of inventions, was as frigid as could be. The Government looked on with a cold and critical eye, and could not be touched either in heart or in pocket.

But with the institution of the Royal Flying Corps the official heart began to warm slightly, and certain tests were laid down for those manufacturers who aspired to sell their machines to the new arm of the Service. These tests, providing for fuel capacity up to 4.0 miles, speeds up to 85 miles an hour, and heights up to 3500 feet, would now be regarded as very elementary affairs. "Looping the loop" was still a dangerous trick for the exhibiting airman and not an evolution; while the "nose–dive" was an uncalculated entry into the next world.

The first important stage in the history of the new arm was reached in July, 1914, when the wing system was abolished, and the Royal Naval Air Service became a separate unit of the Imperial Forces. The first public appearance of the sailor airmen was at a proposed review of the fleet by the King at a test mobilization. The King was unable to attend, but the naval pilots carried out their part of the programme very creditably considering the polyglot nature of their sea–planes. A few weeks later and the country was at war.

There can be no doubt that the Great War has had an enormous forcing influence upon the science of aviation. In times of peace the old game of private enterprise and official neglect would possibly have been carried on in well–marked stages. But with the terrific incentive of victory before them, all Governments fostered the growth of the new arm by all the means in their power. It became a race between Allied and enemy countries as to who first should attain the mastery of the air. The British nation, as usual, started well behind in the race, and their handicap would have been increased to a dangerous extent had Germany not been obsessed by the possibilities of the air–ship as opposed to the aeroplane. Fortunately for us the Zeppelin, as has been described in an earlier chapter, failed to bring about the destruction anticipated by its inventor, and so we gained breathing space for catching up the enemy in the building and equipment of aeroplanes and the training of pilots and observers.

War has set up its usual screens, and the writer is only permitted a very vague and impressionistic picture of the work of the R.F.C. and R.N.A.S. Numerical details and localities must be rigorously suppressed. Descriptions of the work of the Flying Service must be almost as bald as those laconic reports sent in by naval and military airmen to head–quarters. But there is such an accomplishment as reading between the lines.

The flying men fall naturally into two classes—pilots and observers. The latter, of course, act as aerial gunners. The pilots have to pass through three, and observers two, successive courses of training in aviation. Instruction is very detailed and thorough as befits a career which, in addition to embracing the endless problems of flight, demands knowledge of wireless telegraphy, photography, and machine gunnery.

Many of the officers are drafted into the Royal Flying Corps from other branches of the Service, but there are also large numbers of civilians who take up the career. In their case they are first trained as cadets, and, after qualifying for commissions, start their training in aviation at one of the many schools which have now sprung up in all parts of the country.

When the actual flying men are counted in thousands some idea may be gained of the great organization required for the Corps—the schools and flying grounds, the training and activities of the mechanics, the workshops and repair shops, the storage of spare parts, the motor transport, As in other departments of the Service, women have come forward and are doing excellent and most responsible work, especially in the motor–transport section.

A very striking feature of the Corps is the extreme youth of the members, many of the most daring fighters in the air being mere boys of twenty.

The Corps has the very pick of the youth and daring and enterprise of the country. In the days of the old army there existed certain unwritten laws of precedence as between various branches of the Service. If such customs still prevail it is certain that the very newest arm would take pride of place. The flying man has recaptured some of the glamour and romance which encircled the knight–errant of old. He breathes the very atmosphere of dangerous adventure. Life for him is a series of thrills, any one of which would be sufficient to last the ordinary humdrum citizen for a lifetime. Small wonder that the flying man has captured the interest and affection of the people, and all eyes follow these trim, smart, desperadoes of the air in their passage through our cities.

As regards the work of the flying man the danger curve seems to be changing. On the one hand the training is much more severe and exacting than formerly was the case, and so carries a greater element of danger. On the other hand on the battle–front fighting information has in great measure taken the place of the system of men going up "on their own". They are perhaps not so liable to meet with a numerical superiority on the part of enemy machines, which spelt for them almost certain destruction.

For a long time the policy of silence and secrecy which screened "the front" from popular gaze kept us in ignorance of the achievements of our airmen. But finally the voice of the people prevailed in their demand for more enlightenment. Names of regiments began to be mentioned in connection with particular successes. And in the same way the heroes of the R.F.C. and R.N.A.S. were allowed to reap some of the laurels they deserved.

It began to be recognized that publication of the name of an airman who had destroyed a Zeppelin, for instance, did not constitute any vital information to the enemy. In a recent raid upon London the names of the two airmen, Captain G. H. Hackwill, R.F.C., and Lieutenant C. C. Banks, R.F.C., who destroyed a Gotha, were given out in the House of Commons and saluted with cheers. In the old days the secretist party would have regarded this publication as a policy which led the nation in the direct line of "losing the war".

In the annals of the Flying Service, where dare–devilry is taken as a matter of course and hairbreadth escapes from death are part of the daily routine, it is difficult to select adventures for special mention; but the following episodes will give a general idea of the work of the airman in war.

The great feat of Sub–Lieutenant R. A. J. Warneford, R.N.A.S., who single–handed attacked and destroyed a Zeppelin, has already been referred to in Chapter XIII. Lieutenant Warneford was the second on the list of airmen who won the coveted Cross, the first recipient being Second–Lieutenant Barnard Rhodes–Moorhouse, for a daring and successful bomb–dropping raid upon Courtrai in April, 1915. As has happened in so many cases, the award to Lieutenant Rhodes–Moorhouse was a posthumous one, the gallant airman having been mortally wounded during the raid, in spite of which he managed by flying low to reach his destination and make his report.

A writer of adventure stories for boys would be hard put to it to invent any situation more thrilling than that in which Squadron–Commander Richard Bell Davies, D.S.O., R.N., and Flight Sub–Lieutenant Gilbert Formby Smylie, R.N., found themselves while carrying out an air attack upon Ferrijik junction. Smylie's machine was subjected to such heavy fire that it was disabled, and the airman was compelled to plane down after releasing all his bombs but one, which failed to explode. The moment he alighted he set fire to his machine. Presently Smylie saw his companion about to descend quite close to the burning machine. There was infinite danger from the bomb. It was a question of seconds merely before it must explode. So Smylie rushed over to the machine, took hasty aim with his revolver, and exploded the bomb, just before the Commander came within the danger zone. Meanwhile the enemy had commenced to gather round the two airmen, whereupon Squadron–Commander Davies coolly took up the Lieutenant on his machine and flew away with him in safety back to their lines. Davies, who had already won the D.S.O., was given the V.C., while his companion in this amazing adventure was granted the Distinguished Service Cross.

The unexpectedness, to use no stronger term, of life in the R.F.C. in war–time is well exemplified by the adventure which befell Major Rees. The pilot of a "fighter", he saw what he took to be a party of air machines returning from a bombing expedition. Proceeding to join them in the character of escort, Major Rees made the unpleasant discovery that he was just about to join a little party of ten enemy machines. But so far from being dismayed, the plucky airman actually gave battle to the whole ten. One he quickly drove "down and out", as the soldiers say. Attacked by five others, he damaged two of them and dispersed the remainder. Not content with this, he gave chase to two more, and only broke off the engagement when he had received a wound in the thigh. Then he flew home to make the usual laconic report.

No record of heroism in the air could be complete without mention of Captain Ball, who has already figured in these pages. When awarded the V.C. Captain Ball was already the holder of the following honours: D.S.0., M.C., Cross of a Chevalier of the Legion of Honour, and the Russian order of St. George. This heroic boy of twenty was a giant among a company of giants. Here follows the official account which accompanied his award:—

"Lieutenant (temporary Captain) ALBERT BALL, D.S.O., M.C., late Notts and Derby Regiment, and R.F.C.

"For most conspicuous and consistent bravery from April 25 to May 6, 1917, during which period Captain Ball took part in twenty–six combats in the air and destroyed eleven hostile aeroplanes, drove down two out of control, and formed several others to land.

"In these combats Captain Ball, flying alone, on one occasion fought six hostile machines, twice he fought five, and once four.

"While leading two other British aeroplanes he attacked an enemy formation of eight. On each of these occasions he brought down at least one enemy.

"Several times his aeroplane was badly damaged, once so severely that but for the most delicate handling his machine would have collapsed, as nearly all the control wires had been shot away. On returning with a damaged machine, he had always to be restrained from immediately going out on another.

"In all Captain Ball has destroyed forty–three German aeroplanes and one balloon, and has always displayed most exceptional courage, determination, and skill."

So great was Captain Ball's skill as a fighter in the air that for a time he was sent back to England to train new pilots in the schools. But the need for his services at the front was even greater, and it jumped with his desires, for the whole tone of his letters breathes the joy he found in the excitements of flying and fighting. He declares he is having a "topping time", and exults in boyish fashion at a coming presentation to Sir Douglas Haig. It is not too much to say that the whole empire mourned when Captain Ball finally met his death in the air near La Bassee in May, 1917.

CHAPTER XXXIX. Aeroplanes in the Great War

"Aeroplanes and airships would have given us an enormous advantage against the Boers. The difficulty of laying ambushes and traps for isolated columns—a practice at which the enemy were peculiarly adept—would have been very much greater. Some at least of the regrettable reverses which marked the early stages of the campaign could in all probability have been avoided."

So wrote Lord Roberts, our veteran field–marshal, in describing the progress of the Army during recent years. The great soldier was a man who always looked ahead. After his great and strenuous career, instead of taking the rest which he had so thoroughly earned, he spent laborious days travelling up and down the country, warning the people of danger ahead; exhorting them to learn to drill and to shoot; thus attempting to lay the foundation of a great civic army. But his words, alas! fell upon deaf ears—with results so tragic as hardly to bear dwelling upon.

But even "Bobs", seer and true prophet as he was, could hardly have foreseen the swift and dramatic development of war in the air. He had not long been laid to rest when aeroplanes began to be talked about, and, what is more important, to be built, not in hundreds but in thousands. At the time of writing, when we are well into the fourth year of the war, it seems almost impossible for the mind to go back to the old standards, and to take in the statement that the number of machines which accompanied the original Expeditionary Force to France was eighty! Even if one were not entirely ignorant of the number and disposition of the aerial fighting forces over the world–wide battle–ground, the Defence of the Realm Act would prevent us from making public the information. But when, more than a year ago, America entered the war, and talked of building 10,000 aeroplanes, no one gasped. For even in those days one thought of aeroplanes not in hundreds but in tens of thousands.

Before proceeding to give a few details of the most recent work of the Royal Flying Corps and Royal Naval Air Service, mention must be made of the armament of the aeroplane. In the first place, it should be stated that the war has gradually evolved three distinct types of flying machine: (1) the "general–purposes" aeroplane; (2) the giant bomb dropper; (3) the small single–seater "fighter".

As the description implies, the first machine fills a variety of roles, and the duties of its pilots grow more manifold as the war progresses. "Spotting" for the artillery far behind the enemy's lines; "searching" for ammunition dumps, for new dispositions by the enemy of men, material, and guns; attacking a convoy or bodies of troops on the march; sprinkling new trenches with machine–gun fire, or having a go at an aerodrome—any wild form of aerial adventure might be included in the diary of the pilot of a "general–purposes" machine.

It was in order to clear the air for these activities that the "fighter" came into being, and received its baptism of fire at the Battle of the Somme. At first the idea of a machine for fighting only, was ridiculed. Even the Germans, who, in a military sense, were awake and plotting when other nations were dozing in the sunshine of peace, did not think ahead and imagine the aerial duel between groups of aeroplanes armed with machine–guns. But soon the mastery of the air became of paramount importance, and so the fighter was evolved. Nobly, too, did the men of all nations rise to these heroic and dangerous opportunities. The Germans were the first to boast of the exploits of their fighting airmen, and to us in Britain the names of Immelmann and Bolcke were known long before those of any of our own fighters. The former claimed not far short of a hundred victims before he was at last brought low in June, 1916. His letters to his family were published soon after his death, and do not err on the side of modesty.

On 11th August, 1915, he writes: "There is not much doing here. Ten minutes after Bolcke and I go up, there is not an enemy airman to be seen. The English seem to have lost all pleasure in flying. They come over very, very seldom."

When allowance has been made for German brag, these statements throw some light upon the standard of British flying at a comparatively early date in the war. Certainly no German airman could have made any such complaint a year later. In 1917 the German airmen were given all the fighting they required and a bit over.

Certainly a very different picture is presented by the dismal letters which Fritz sent home during the great Ypres offensive of August, 1917. In these letters he bewails the fact that one after another of his batteries is put out of action owing to the perfect "spotting" of the British airmen, and arrives at the sad conclusion that Germany has lost her superiority in the air.

An account has already been given of the skill and prowess of Captain Ball. On his own count—and he was not the type of man to exaggerate his prowess—he found he had destroyed fifty machines, although actually he got the credit for forty–one. This slight discrepancy may be explained by the scrupulous care which is taken to check the official returns. The air fighter, though morally certain of the destruction of a certain enemy aeroplane, has to bring independent witnesses to substantiate his claim, and when out "on his own" this is no easy matter. Without this check, though occasionally it acts harshly towards the pilot, there might be a tendency to exaggerate enemy losses, owing to the difficulty of distinguishing between an aeroplane put out of action and one the pilot of which takes a sensational "nose dive" to get out of danger.

One of the most striking illustrations of the growth of the aeroplane as a fighting force is afforded by the great increase in the heights at which they could scout, take photographs, and fight. In Sir John French's dispatches mention is made of bomb–dropping from 3000 feet. In these days the aerial battleground has been extended to anything up to 20,000 feet. Indeed, so brisk has been the duel between gun and aeroplane, that nowadays airmen have often to seek the other margin of safety, and can defy the anti–aircraft guns only by flying so low as just to escape the ground. The general armament of a "fighter" consists of a maxim firing through the propeller, and a Lewis gun at the rear on a revolving gun–ring.

It is pleasant to record that the Allies kept well ahead of the enemy in their use of aerial photography. Before a great offensive some thousands of photographs had to be taken of enemy dispositions by means of cameras built into the aeroplanes.

Plates were found to stand the rough usage better than films, and not for the first time in the history of mechanics the man beat the machine, a skilful operator being found superior to the ingenious automatic plate–fillers which had been devised.

The counter–measure to this ruthless exposure of plans was camouflage. As if by magic–tents, huts, dumps, guns began, as it were, to sink into the scenery. The magicians were men skilled in the use of brush and paint–pot, and several leading figures in the world of art lent their services to the military authorities as directors of this campaign of concealment. In this connection it is interesting to note that both Admiralty and War Office took measures to record the pictorial side of the Great War. Special commissions were given to a notable band of artists working in their different "lines". An abiding record of the great struggle will be afforded by the black–and–white work of Muirhead Bone, James M'Bey, and Charles Pears; the portraits, landscapes, and seascapes of Sir John Lavery, Philip Connard, Norman Wilkinson, and Augustus John, who received his commission from the Canadian Government.

CHAPTER XL. The Atmosphere and the Barometer

For the discovery of how to find the atmospheric pressure we are indebted to an Italian named Torricelli, a pupil of Galileo, who carried out numerous experiments on the atmosphere toward the close of the sixteenth century.

Torricelli argued that, as air is a fluid, if it had weight it could be made to balance another fluid of known weight. In his experiments he found that if a glass tube about 3 feet in length, open at one end only, and filled with mercury, were placed vertically with the open end submerged in a cup of mercury, some of the mercury in the tube descended into the cup, leaving a column of mercury about 30 inches in height in the tube. From this it was deduced that the pressure of air on the surface of the mercury in the cup forced it up the tube to the height Of 30 inches, and this was so because the weight of a column of air from the cup to the top of the atmosphere was only equal to that of a column of mercury of the same base and 30 inches high.

Torricelli's experiment can be easily repeated. Take a glass tube about 3 feet long, closed at one end and open at the other; fill it as full as possible with mercury. Then close the open end with the thumb, and invert the tube in a basin of mercury so that the open end dips beneath the surface. The mercury in the tube will be found to fall a short distance, and if the height of the column from the surface of the mercury in the basin be measured you will find it will be about 30 inches. As the tube is closed at the top there is no downward pressure of air at that point, and the space above the mercury in the tube is quite empty: it forms a VACUUM. This vacuum is generally known as the TORRICELLIAN VACUUM, after the name of its discoverer.

Suppose, now, a hole be bored through the top of the tube above the column of mercury, the mercury will immediately fall in the tube until it stands at the same level as the mercury in the basin, because the upward pressure of air through the liquid in the basin would be counterbalanced by the downward pressure of the air at the top, and the mercury would fall by its own weight.

A few years later Professor Boyle proposed to use the instrument to measure the height of mountains. He argued that, since the pressure of the atmosphere balanced a column of mercury 30 inches high, it followed that if one could find the weight of the mercury column one would also find the weight of a column of air standing on a base of the same size, and stretching away indefinitely into space. It was found that a column of mercury in a tube having a sectional area of 1 square inch, and a height of 30 inches, weighed 15 pounds; therefore the weight of the atmosphere, or air pressure, at sea–level is about 15 pounds to the square inch. The ordinary mercury barometer is essentially a Torricellian tube graduated so that the varying heights of the mercury column can be used as a measure of the varying atmospheric pressure due to change of weather or due to alteration of altitude. If we take a mercury barometer up a hill we will observe that the mercury falls. The weight of atmosphere being less as we ascend, the column of mercury supported becomes smaller.

Although the atmosphere has been proved to be over 200 miles high, it has by no means the same density throughout. Like all gases, air is subject to the law that the density increases directly as the pressure, and thus the densest and heaviest layers are those nearest the sea–level, because the air near the earth's surface has to support the pressure of all the air above it. As airmen rise into the highest portions of the atmosphere the height of the column of air above them decreases, and it follows that, having a shorter column of air to support, those portions are less dense than those lower down. So rare does the atmosphere become, when great altitudes are reached, that at a height of seven miles breathing is well–nigh impossible, and at far lower altitudes than this airmen have to be supported by inhalations of oxygen.

One of the greatest altitudes was reached by two famous balloonists, Messrs. Coxwell and Glaisher. They were over seven miles in the air when the latter fell unconscious, and the plucky aeronauts were only saved by Mr. Coxwell pulling the valve line with his teeth, as all his limbs were disabled.

CHAPTER XLI. How an Airman Knows what Height he Reaches

One of the first questions the visitor to an aerodrome, when watching the altitude tests, asks is: "How is it known that the airman has risen to a height of so many feet?" Does he guess at the distance he is above the earth?

If this were so, then it is very evident that there would be great difficulty in awarding a prize to a number of competitors each trying to ascend higher than his rivals.

No; the pilot does not guess at his flying height, but he finds it by a height–recording instrument called the BAROGRAPH.

In the last chapter we saw how the ordinary mercurial barometer can be used to ascertain fairly accurately the height of mountains. But the airman does not take a mercurial barometer up with him. There is for his use another form of barometer much more suited to his purpose, namely, the barograph, which is really a development of the aneroid barometer.

The aneroid barometer (Gr. a, not; neros, moist) is so called because it requires neither mercury, glycerine, water, nor any other liquid in its construction. It consists essentially of a small, flat, metallic box made of elastic metal, and from which the air has been partially exhausted. In the interior there is an ingenious arrangement of springs and levers, which respond to atmospheric pressure, and the depression or elevation of the surface is registered by an index on the dial. As the pressure of the atmosphere increases, the sides of the box are squeezed in by the weight of the air, while with a decrease of pressure they are pressed out again by the springs. By means of a suitable adjustment the pointer on the dial responds to these movements. It is moved in one direction for increase of air pressure, and in the opposite for decreased pressure. The positions of the figures on the dial are originally obtained by numerous comparisons with a standard mercurial barometer, and the scale is graduated to correspond with the mercurial barometer.

From the illustration here given you will notice the pointer and scale of the "A. G" aero–barograph, which is used by many of our leading airmen, and which, as we have said, is a development of the aneroid barometer. The need of a self–registering scale to a pilot who is competing in an altitude test, or who is trying to establish a height record, is self–evident. He need not interfere with the instrument in the slightest; it records and tells its own story. There is in use a pocket barograph which weighs only 1 pound, and registers up to 4000 feet.

It is claimed for the "A. G." barograph that it is the most precise instrument of its kind. Its advantages are that it is quite portable—it measures only 6 1/4 inches in length, 3 1/2 inches in width, and 2 1/2 inches in depth, with a total weight of only 14 pounds—and that it is exceptionally accurate and strong. Some idea of the labour involved in its construction may be gathered from the fact that this small and insignificant–looking instrument, fitted in its aluminium case, costs over L8.

CHAPTER XLII. How an Airman finds his Way

In the early days of aviation we frequently heard of an aviator losing his way, and being compelled to descend some miles from his required destination. There are on record various instances where airmen have lost their way when flying over the sea, and have drifted so far from land that they have been drowned. One of the most notable of such disasters was that which occurred to Mr. Hamel in 1914, when he was trying to cross the

English Channel. It is presumed that this unfortunate pilot lost his bearings in a fog, and that an, accident to his machine, or a shortage of petrol, caused him to fall in the sea.

There are several reasons why air pilots go out of their course, even though they are supplied with most efficient compasses. One cause of misdirection is the prevalence of a strong side wind. Suppose, for example, an airman intended to fly from Harwich to Amsterdam. A glance at the map will show that the latter place is almost due east of Harwich. We will assume that when the pilot leaves Earth at Harwich the wind is blowing to the east; that is, behind his back.

Now, however strong a wind may be, and in whatever direction it blows, it always appears to be blowing full in a pilot's face. Of course this is due to the fact that the rush of the machine through the air "makes a wind", as we say. Much the same sort of thing is experienced on a bicycle; when out cycling we very generally seem to have a "head" wind.

Suppose during his journey a very strong side wind sprang,up over the North Sea. The pilot would still keep steering his craft due east, and it must be remembered that when well out at sea there would be no familiar landmarks to guide him, so that he would have to rely solely on his compass. It is highly probable that he would not feel the change of wind at all, but it is even more probable that when land was ultimately reached he would be dozens of miles from his required landing–place.

Quite recently Mr. Alexander Gross, the well–known maker of aviation instruments, who is even more famous for his excellent aviation maps, claims to have produced an anti–drift aero–compass, which has been specially designed for use on aeroplanes. The chief advantages of this compass are that the dial is absolutely steady; the needle is extremely sensitive and shows accurately the most minute change of course; the anti–drift arrangement checks the slightest deviation from the straight course; and it is fitted with a revolving sighting arrangement which is of great importance in the adjustment of the instrument.

Before the airman leaves Earth he sets his compass to the course to be steered, and during the flight he has only to see that the two boldly–marked north points—on the dial and on the outer ring—coincide to know that he is keeping his course. The north points are luminous, so that they are clearly visible at night.

It is quite possible that if some of our early aviators had carried such a highly–efficient compass as this, their lives might have been saved, for they would not have gone so far astray in their course. The anti–drift compass has been adopted by various Governments, and it now forms part of the equipment of the Austrian military aeroplane.

When undertaking cross–country flights over strange land an airman finds his way by a specially–prepared map which is spread out before him in an aluminium map case. From the illustration here given of an aviator's map, you will see that it differs in many respects from the ordinary map. Most British aviation maps are made and supplied by Mr Alexander Gross, of the firm of "Geographia", London.

Many airmen seem to find their way instinctively, so to speak, and some are much better in picking out landmarks, and recognizing the country generally, than others. This is the case even with pedestrians, who have the guidance of sign–posts, street names, and so on to assist them. However accurately some people are directed, they appear to have the greatest difficulty in finding their way, while others, more fortunate, remember prominent features on the route, and pick out their course as accurately as does a homing pigeon.

Large sheets of water form admirable "sign–posts" for an airman; thus at Kempton Park, one of the turning–points in the course followed in the "Aerial Derby", there are large reservoirs, which enable the airmen to follow the course at this point with the greatest ease. Railway lines, forests, rivers and canals, large towns, prominent structures, such as gasholders, chimney–stalks, and so on, all assist an airman to find his way.

CHAPTER XLIII. The First Airman to Fly Upside Down

Visitors to Brooklands aerodrome on 25th September, 1913, saw one of the greatest sensations in this or any other century, for on that date a daring French aviator, M. Pegoud, performed the hazardous feat of flying upside down.

Before we describe the marvellous somersaults which Pegoud made, two or three thousand feet above the earth, it would be well to see what was the practical use of it all. If this amazing airman had been performing some circus trick in the air simply for the sake of attracting large crowds of people to witness it, and therefore being the means of bringing great monetary gain both to him and his patrons, then this chapter would never have been written. Indeed, such a risk to one's life, if there had been nothing to learn from it, would have been foolish.

No; Pegoud's thrilling performance must be looked at from an entirely different standpoint to such feats of daring as the placing of one's head in the jaws of a lion, the traversing of Niagara Falls by means of a tight–rope stretched across them, and other similar senseless acts, which are utterly useless to mankind.

Let us see what such a celebrated airman as Mr. Gustav Hamel said of the pioneer of upside–down flying.

"His looping the loop, his upside–down flights, his general acrobatic feats in the air are all of the utmost value to pilots throughout the world. We shall have proof of this, I am sure, in the near future. Pegoud has shown us what it is possible to do with a modern machine. In his first attempt to fly upside down he courted death. Like all pioneers, he was taking liberties with the unknown elements. No man before him had attempted the feat. It is true that men have been upside down in the air; but they were turned over by sudden gusts of wind, and in most cases were killed. Pegoud is all the time rehearsing accidents and showing how easy it is for a pilot to recover equilibrium providing he remains perfectly calm and clear–headed. Any one of his extraordinary positions might be brought about by adverse elements. It is quite conceivable that a sudden gust of wind might turn the machine completely over. Hitherto any pilot in such circumstances would give himself up for lost. Pegoud has taught us what to do in such a case. . . . his flights have given us all a new confidence.

"In a gale the machine might be upset at many different angles. Pegoud has shown us that it is easily possible to recover from such predicaments. He has dealt with nearly every kind of awkward position into which one might be driven in a gale of wind, or in a flight over mountains where air–currents prevail.

"He has thus gained evidence which will be of the utmost value to present and future pilots, and prove a factor of signal importance in the preservation of life in the air."

Such words as these, coming from a man of Mr. Hamel's reputation as an aviator, clearly show us that M. Pegoud has a life–saving mission for airmen throughout the world.

Let us stand, in imagination, with the enormous crowd of spectators who invaded the Surrey aerodrome on 25th September, and the two following days, in 1913.

What an enormous crowd it was! A line of motor–cars bordered the track for half a mile, and many of the spectators were busy city men who had taken a hasty lunch and rushed off down to Weybridge to see a little French airman risk his life in the air. Thousands of foot passengers toiled along the dusty road from the paddock to the hangars, and thousands more, who did not care to pay the shilling entrance fee, stood closely packed on the high ground outside the aerodrome.

Biplanes and monoplanes came driving through the air from Hendon, and airmen of world–wide fame, such as Sopwith, Hamel, Verrier, and Hucks, had gathered together as disciples of the great life–saving missionary. Stern critics these! Men who would ruthlessly expose any "faked" performance if need were!

And where is the little airman while all this crowd is gathering? Is he very excited? He has never before been in England. We wonder if his amazing coolness and admirable control over his nerves will desert him among strange surroundings.

Probably Pegoud was the coolest man in all that vast crowd. He seemed to want to hide himself from public gaze. Most of his time, was taken up in signing post–cards for people who had been fortunate enough to discover him in a little restaurant near which his shed was situated.

At last his Bleriot monoplane was wheeled out, and he was strapped, or harnessed, into his seat. "Was the machine a 'freak' monoplane?" we wondered.

We were soon assured that such was not the case. Indeed, as Pegoud himself says: "I have used a standard type of monoplane on purpose. Almost every aeroplane, if it is properly balanced, has just as good a chance as mine, and I lay particular stress on the fact that there is nothing extraordinary about my machine, so that no one can say my achievements are in any way faked."

During the preliminary operations his patron, M. Bleriot, stood beside the machine, and chatted affably with the aviator. At last the signal was given for his ascent, and in a few moments Pegoud was climbing with the nose of his machine tilted high in the air. For about a quarter of an hour he flew round in ever–widening circles, rising very quietly and steadily until he had reached an altitude of about 4000 feet. A deep silence seemed to have settled on the vast crowd nearly a mile below, and the musical droning of his engine could be plainly heard.

Then his movements began to be eccentric. First, he gave a wonderful exhibition of banking at right angles. Then, after about ten minutes, he shut off his engine, pitched downwards and gracefully righted himself again.

At last the amazing feat began. His left wing was raised, his right wing dipped, and the nose of the machine dived steeply, and turned right round with the airman hanging head downwards, and the wheels of the monoplane uppermost. In this way he travelled for about a hundred yards, and then slowly righted the machine, until it assumed its normal position, with the engine again running. Twice more the performance was repeated, so that he travelled from one side of the aerodrome to the other—a distance of about a mile and a half.

Next he descended from 4000 feet to about 1200 feet in four gigantic loops, and, as one writer said: "He was doing exactly what the clown in the pantomime does when he climbs to the top of a staircase and rolls deliberately over and over until he reaches the ground. But this funny man stopped before he reached the ground, and took his last flight as gracefully as a Columbine with outspread skirts."

Time after time Pegoud made a series of S–shaped dives, somersaults, and spiral descents, until, after an exhibition which thrilled quite 50,000 people, he planed gently to Earth.

Hitherto Pegoud's somersaults have been made by turning over from front to back, but the daring aviator, and others who followed him, afterwards turned over from right to left or from left to right. Pegoud claimed to have demonstrated that the aeroplane is uncapsizeable, and that if a parachute be attached to the fuselage, which is the equivalent of a life boat on board a ship, then every pilot should feel as safe in a heavier–than–air machine as in a motor–car.

CHAPTER XLIV. The First Englishman to Fly Upside Down

After M. Pegoud's exhibition of upside–down flying in this country it was only to be expected that British aviators would emulate his daring feat. Indeed, on the same day that the little Frenchman was turning somersaults in the air at Brooklands Mr. Hamel was asking M. Bleriot for a machine similar to that used by Pegoud, so that he might demonstrate to airmen the stability of the aeroplane in almost all conceivable positions.

However, it was not the daring and skilful Hamel who had the honour of first following in Pegoud's footsteps, but another celebrated pilot, Mr. Hucks.

Mr. Hucks was an interested spectator at Brooklands when Pegoud flew there in September, and he felt that, given similar conditions, there was no reason why he should not repeat Pegoud's performance. He therefore talked the matter over with M. Bleriot, and began practising for his great ordeal.

His first feat was to hang upside–down in a chair supported by a beam in one of the sheds, so that he would gradually become accustomed to the novel position. For a time this was not at all easy. Have you ever tried to stand on your hands with your feet upwards for any length of time? To realize the difficulty of being head downwards, just do this, and get someone to hold your legs. The blood will, of course, "rush to the head", as we say, and the congestion of the blood–vessels in this part of the body will make you feel extremely dizzy. Such an occurrence would be fatal in an aeroplane nearly a mile high in the air at a time when one requires an especially clear brain to manipulate the various controls.

But, strange to say, the airman gradually became used to the "heels–over–head" position, and, feeling sure of himself, he determined to start on his perilous undertaking. No one with the exception of M. Bleriot and the mechanics were present at the Buc aerodrome, near Versailles, when Mr. Hucks had his monoplane brought out with the intention of looping the loop.

He quickly rose to a height of 1500 feet, and then, slowly dipping the nose of his machine, turned right over. For fully half a minute he flew underneath the monoplane, and then gradually brought it round to the normal position.

In the afternoon he continued his experiments, but this time at a height of nearly 3000 feet. At this altitude he was flying quite steadily, when suddenly he assumed a perpendicular position, and made a dive of about 600 feet. The horrified spectators thought that the gallant aviator had lost control of his machine and was dashing straight to Earth, but quickly he changed his direction and slowly planed upwards. Then almost as suddenly he turned a complete somersault. Righting the aeroplane, he rose in a succession of spiral flights to a height of between 3000 and 3500 feet, and then looped the loop twice in quick succession.

On coming to earth M. Bleriot heartily congratulated the brave Englishman. Mr. Hucks admitted a little nervousness before looping the loop; but, as he remarked: "Once I started to go round my nervousness vanished, and then I knew I was coming out on top. It is all a question of keeping control of your nerves, and Pegoud deserved all the credit, for he was the first to risk his life in flying head downwards."

Mr. Hucks intended to be the first Englishman to fly upside down in England, but he was forestalled by one of our youngest airmen, Mr. George Lee Temple. On account of his youth—Mr. Temple was only twenty-one at the time when he first flew upside-down—he was known as the "baby airman", but there was probably no more plucky airman in the world.

There were special difficulties which Mr. Temple had to overcome that did not exist in the experiments of M. Pegoud or Mr. Hucks. To start with, his machine—a 50-horse-power Bleriot monoplane—was said by the makers to be unsuitable for the performance. Then he could get no assistance from the big aeroplane firms, who sought to dissuade him from his hazardous undertaking. Experienced aviators wisely shook their heads and told the "baby airman" that he should have more practice before he took such a risk.

But notwithstanding this lack of encouragement he practised hard for a few days by hanging in an inverted position. Meanwhile his mechanics were working night and day in strengthening the wings of the monoplane, and fitting it with a slightly larger elevator.

On 24th November, 1913, he decided to "try his luck" at the London aerodrome. He was harnessed into his seat, and, bidding his friends farewell, with the words "wish me luck", he went aloft. For nearly half an hour he climbed upward, and swooped over the aerodrome in wide circles, while his friends far below were watching every action of his machine.

Suddenly an alarming incident occurred. When about a mile high in the air the machine tipped downwards and rushed towards Earth at terrific speed. Then the tail of the machine came up, and the "baby airman" was hanging head downwards.

But at this point the group of airmen standing in the aerodrome were filled with alarm, for it was quite evident to their experienced eyes that the monoplane was not under proper control. Indeed, it was actually side–slipping, and a terrible disaster appeared imminent. For hundreds of feet the young pilot, still hanging head downwards, was crashing to Earth, but when down to about 1200 feet from the ground the machine gradually came round, and Mr. Temple descended safely to Earth.

The airman afterwards told his friends that for several seconds he could not get the machine to answer the controls, and for a time he was falling at a speed of 100 miles an hour. In ordinary circumstances he thought that a dive of 500 feet after the upside–down stretch should get him the right way up, but it really took him nearly 1500 feet. Fortunately, however, he commenced the dive at a great altitude, and so the distance side–slipped did not much matter.

It is sad to relate that Mr. Temple lost his life in January, 1914, while flying at Hendon in a treacherous wind. The actual cause of the accident was never clearly understood. He had not fully recovered from an attack of influenza, and it was thought that he fainted and fell over the control lever while descending near one of the pylons, when the machine "turned turtle", and the pilot's neck was broken.

CHAPTER XLV. Accidents and their Cause

"Another airman killed!" "There'll soon be none of those flying fellows left!" "Far too risky a game!" "Ought to be stopped by law!"

How many times have we heard these, and similar remarks, when the newspapers relate the account of some fatality in the air! People have come to think that flying is a terribly risky occupation, and that if one wishes to put an end to one's life one has only to go up in a flying machine. For the last twenty years some of our great writers have prophesied that the conquest of the air would be as costly in human life as was that of the sea, but their prophecies have most certainly been wrong, for in the wreck of one single vessel, such as that of the Titanic, more lives were lost than in all the disasters to any form of aerial craft.

Perhaps some of our grandfathers can remember the dread with which many nervous people entered, or saw their friends enter, a train. Travellers by the railway eighty or ninety years ago considered that they took their lives in their hands, so to speak, when they went on a long journey, and a great sigh of relief arose in the members of their families when the news came that the journey was safely ended. In George Stephenson's days there was considerable opposition to the institution of the railway, simply on account of the number of accidents which it was anticipated would take place.

Now we laugh at the fears of our great–grandparents; is it not probable that our grandchildren will laugh in a similar manner at our timidity over the aeroplane?

In the case of all recent new inventions in methods of locomotion there has always been a feeling among certain people that the law ought to prohibit such inventions from being put into practice.

There used to be bitter opposition to the motor–car, and at first every mechanically–driven vehicle had to have a man walking in front with a red flag.

There are risks in all means of transit; indeed, it may be said that the world is a dangerous place to live in. It is true, too, that the demons of the air have taken their toll of life from the young, ambitious, and daring souls. Many of the fatal accidents have been due to defective work in some part of the machinery, some to want of that complete knowledge and control that only experience can give, some even to want of proper care on the part of the pilot. If a pilot takes ordinary care in controlling his machine, and if the mechanics who have built the machine have done their work thoroughly, flying, nowadays, should be practically as safe as motoring.

The French Aero Club find, from a mass or information which has been compiled for them with great care, that for every 92,000 miles actually flown by aeroplane during the year 1912, only one fatal accident had occurred. This, too, in France, where some of the pilots have been notoriously reckless, and where far more airmen have been killed than in Britain.

When we examine carefully the statistics dealing with fatal accidents in aeroplanes we find that the pioneers of flying, such as the famous Wright Brothers, Bleriot, Farman, Grahame–White, and so on, were comparatively free from accidents. No doubt, in some cases, defective machines or treacherous wind gusts caused the craft to collapse in mid–air. But, as a rule, the first men to fly were careful to see that every part of the machine was in order before going up in it, so that they rarely came to grief through the planes not being sufficiently tightened up, wires being unduly strained, spars snapping, or bolts becoming loose.

Mr. Grahame–White admirably expresses this when he says: "It is a melancholy reflection, when one is going through the lists of aeroplane fatalities, to think how many might have been avoided. Really the crux of the situation in this connection, as it appears to me, is this: the first men who flew, having had all the drudgery and danger of pioneer work, were extremely careful in all they did; and this fact accounts for the comparatively large proportion of these very first airmen who have survived.

"But the men who came next in the path of progress, having a machine ready–made, so to speak, and having nothing to do but to get into it and fly, did not, in many cases, exercise this saving grace of caution. And that—at least in my view—is why a good many of what one may call the second flight of pilots came to grief."

CHAPTER XLVI. Accidents and their Cause (Cont.)

One of the main causes of aeroplane accidents has been the breakage of some part of the machine while in the air, due to defective work in its construction. There is no doubt that air–craft are far more trustworthy now than they were two or three years ago. Builders have learned from the mistakes of their predecessors as well as profited by their own. After every serious accident there is an official enquiry as to the probable cause of the accident, and information of inestimable value has been obtained from such enquiries.

The Royal Aero Club of Great Britain has a special "Accidents Investigation Committee" whose duty it is to issue a full report on every fatal accident which occurs to an aeroplane in this country. As a rule, representatives of the committee visit the scene of the accident as soon as possible after its occurrence. Eye–witnesses are called before them to give evidence of the disaster; the remains of the craft are carefully inspected in order to discover any flaw in its construction; evidence is taken as to the nature and velocity of the wind on the day of the accident, the approximate height at which the aviator was flying, and, in fact, everything of value that might bear on the cause of the accident.

As a good example of an official report we may quote that issued by the Accidents Investigation Committee of the Royal Aero Club on the fatal accident which occurred to Colonel Cody and his passenger on 7th August, 1913.

"The representatives of the Accidents Committee visited the scene of the accident within a few hours of its occurrence, and made a careful examination of the wrecked air–craft. Evidence was also taken from the eye–witnesses of the accident.

"From the consideration of the evidence the Committee regards the following facts as clearly established:

"1. The air–craft was built at Farnborough, by Mr. S. F. Cody, in July, 1913.

"2. It was a new type, designed for the Daily Mail Hydroplane Race round Great Britain, but at the time of the accident had a land chassis instead of floats.

"3. The wind at the time of the accident was about 10 miles per hour.

"4. At about 200 feet from the ground the air–craft buckled up and fell to the ground. A large piece of the lower left wing, composing the whole of the front spar between the fuselage and the first upright, was picked up at least 100 yards from the spot where the air–craft struck the ground.

"5. The fall of the air–craft was broken considerably by the trees, to such an extent that the portion of the fuselage surrounding the seats was practically undamaged.

"6. Neither the pilot nor passenger was strapped in.

"Opinion. The Committee is of opinion that the failure of the air–craft was due to inherent structural weakness.

"Since that portion of the air–craft in which the pilot and passenger were seated was undamaged, it is conceivable their lives might have been saved had they been strapped in."

This occasion was not the only time when the Accidents Investigation Committee recommended the advisability of the airman being strapped to his seat. But many airmen absolutely refuse to wear a belt, just as many cyclists cannot bear to have their feet made fast to the pedals of their cycles by using toe–clips.

Mention of toe–clips brings us to other accidents which sometimes befall airmen. As we have seen in a previous chapter, Mr. Hawker's accident in Ireland was due to his foot slipping over the rudder bar of his machine. It is thought that the disaster to Mr. Pickles' machine on "Aerial Derby" day in 1913 was due to the same cause, and on one occasion Mr. Brock was in great danger through his foot slipping on the rudder bar while he was practising some evolutions at the London Aerodome. Machines are generally flying at a very fast rate, and if the pilot loses control of the machine when it is near the ground the chances are that the aeroplane crashes to earth before he can right it. Both Mr. Hawker and Mr. Pickles were flying low at the time of their accidents, and so their machines were smashed; fortunately Mr. Brock was comparatively high up in the air, and though his machine rocked about and banked in an ominous manner, yet he was able to gain control just in the nick of time.

To prevent accidents of this kind the rudder bars could be fitted with pedals to which the pilot's feet could be secured by toe–clips, as on bicycle pedals. Indeed, some makers of air–craft have already provided pedals with toe–clips for the rudder bar. Probably some safety device such as this will soon be made compulsory on all machines.

We have already remarked that certain pilots do not pay sufficient heed to the inspection of their machines before making a flight. The difference between pilots in this respect is interesting to observe. On the great day at Hendon, in 1913—the Aerial Derby day—there were over a dozen pilots out with their craft.

From the enclosure one could watch the airmen and their mechanics as the machines were run out from the hangars on to the flying ground. One pilot walked beside his mechanics while they were running the machine to the starting place, and watched his craft with almost fatherly interest. Before climbing into his seat he would carefully inspect the spars, bolts, wires, controls, and so on; then he would adjust his helmet and fasten himself into his seat with a safety belt.

"Surely with all that preliminary work he is ready to start," remarked one of the spectators standing by. But no! the engine must be run at varying speeds, while the mechanics hold back the machine. This operation alone took three or four minutes, and all that the pilot proposed to do was to circle the aerodrome two or three times. An onlooker asked a mechanic if there were anything wrong with that particular machine. "No!" was the reply; "but our governor's very faddy, you know!"

And now for the other extreme! Three mechanics emerged from a hangar pushing a rather ungainly–looking biplane, which bumped over the uneven ground. The pilot was some distance behind, with cigarette in mouth, joking with two or three friends. When the machine was run out into the open ground he skipped quickly up to it, climbed into the seat, started the engine, waved a smiling "good–bye", and was off. For all he knew, that rather rough jolting of the craft while it was being removed from the hangar might have broken some wire on which the safety of his machine, and his life, depended. The excuse cannot be made that his mechanics had performed this all–important work of inspection, for their attention was centred on the daring "banking " evolutions of some audacious pilot in the aerodrome.

Mr. C. G. Grey, the well–known writer on aviation matters, and the editor of The Aeroplane, says, with regard to the need of inspection of air–craft:—

"A pilot is simply asking for trouble if he does not go all over his machine himself at least once a day, and, if possible, every time he is starting for a flight.

"One seldom hears, in these days, of a broken wheel or axle on a railway coach, yet at the chief stopping places on our railways a man goes round each train as it comes in, tapping the tires with a hammer to detect cracks, feeling the hubs to see if there is any sign of a hot box, and looking into the grease containers to see if there is a proper supply of lubricant. There ought to be a similar inspection of every aeroplane every time it touches the ground. The jar of even the best of landings may fracture a bolt holding a wire, so that when the machine goes up again the wire may fly back and break the propeller, or get tangled in the control wires, or a strut or socket may crack in landing, and many other things may happen which careful inspection would disclose before any harm could occur. Mechanics who inspected machines regularly would be able to go all over them in a few minutes, and no time would be wasted. As it is, at any aerodrome one sees a machine come down, the pilot and passenger (a fare or a pupil) climb out, the mechanics hang round and smoke cigarettes, unless they have to perform the arduous duties of filling up with petrol. In due course another passenger and a pilot climb in, a mechanic swings the propeller, and away they go quite happily. If anything casts loose they come down—and it is truly wonderful how many things can come loose or break in the air without anyone being killed. If some thing breaks in landing, and does not actually fall out of place, it is simply a matter of luck whether anyone happens to see it or not."

This advice, coming from a man with such wide experience of the theory and practice of flying, should surely be heeded by all those who engage in deadly combat with the demons of the air. In the early days of aviation, pilots were unacquainted with the nature and method of approach of treacherous wind gusts; often when they were flying along in a steady, regular wind, one of these gusts would strike their craft on one side, and either overturn it or cause it to over–bank, so that it crashed to earth with a swift side–slip through the air.

Happily the experience of those days, though purchased at the cost of many lives, has taught makers of air–craft to design their machines on more trustworthy lines. Pilots, too, have made a scientific study of air eddies, gusts, and so on, and the danger of flying in a strong or gusty wind is comparatively small.

CHAPTER XLVII. Accidents and their Cause (Cont.)

Many people still think that if the engine of an aeroplane should stop while the machine was in mid–air, a terrible disaster would happen. All petrol engines may be described as fickle in their behaviour, and so complicated is their structure that the best of them are given to stopping without any warning. Aeroplane engines are far superior in horse–power to those fitted to motorcars, and consequently their structure is more intricate. But if an airman's engine suddenly stopped there would be no reason whatever why he should tumble down head first and break his neck. Strange to say, too, the higher he was flying the safer he would be.

All machines have what is called a GLIDING ANGLE. When the designer plans his machine he considers the distribution of the weight or the engine, pilot and passengers, of the petrol, aeronautical instruments, and planes, so that the aeroplane is built in such a manner that when the engine stops, and the nose of the machine is turned downwards, the aeroplane of its own accord takes up its gliding angle and glides to earth.

Gliding angles vary in different machines. If the angle is one in twelve, this would mean that if the glide wave commenced at a height of 1 mile, and continued in a straight line, the pilot would come to earth 12 miles distant. We are all familiar with the gradients shown on railways. There we see displayed on short sign–posts such notices as "1 in 50", with the opposite arms of the post pointing upwards and downwards. This, of course, means that the slope of the railway at that particular place is 1 foot in a distance of 50 feet.

One in twelve may be described as the natural gradient which the machine automatically makes when engine power is cut off. It will be evident why it is safer for a pilot to fly, say, at four or five thousand feet high than just over the tree–tops or the chimney–pots of towns. Suppose, for example, the machine has a gliding angle of one in twelve, and that when at an altitude of about a mile the engine should stop. We will assume that at the time of the stoppage the pilot is over a forest where it is quite impossible to land. Directly the engine stopped he would change the angle of the elevating plane, so that the aeroplane would naturally fall into its gliding angle. The craft would at once settle itself into a forward and slightly downward glide; and the airman, from his point of vantage, would be able to see the extent of the forest. We will assume that the aeroplane is gliding in a northerly direction, and that the country is almost as unfavourable for landing there as over the forest itself. In fact, we will imagine an extreme case, where the airman is over country quite unsuitable for landing except toward the south; that is, exactly opposite to the direction in which he starts to glide. Fortunately, there is no reason why he should not steer his machine right round in the air, even though the only power is that derived from the force of gravity. His descent would be in an immense slope, extending 10 or 12 miles from the place where the engine stopped working. He would therefore be able to choose a suitable landing–place and reach earth quite safely.

But supposing the airman to be flying about a hundred yards above the forest–an occurrence not likely to happen with a skilled airman, who would probably take an altitude of nearly a mile. Almost before he could have time to alter his elevating plane, and certainly long before he could reach open ground, he would be on the tree–tops.

It is thought that in the near future air–craft will, be fitted with two or more motors, so that when one fails the other will keep the machine on its course. This has been found necessary in Zeppelin air–ships. In an early Zeppelin model, which was provided with one engine only, the insufficient power caused the pilot to descend on unfavourable ground, and his vessel was wrecked. More recent types of Zeppelins are fitted with three or four engines. Experiments have already been made with the dual–engine plant for aeroplanes, notably by Messrs. Short Brothers, of Rochester, and the tests have given every satisfaction.

There is little doubt that if the large passenger aeroplane is made possible, and if parliamentary powers have to be obtained for the formation of companies for passenger traffic by aeroplane, it will be made compulsory to fit machines with two or more engines, driving three or four distinct propellers. One of the engines would possibly be of inferior power, and used only in cases of emergency.

Still another cause of accident, which in some cases has proved fatal, is the taking of unnecessary risks when in the air. This has happened more in America and in France than in Great Britain. An airman may have performed a very difficult and daring feat at some flying exhibition and the papers belauded his courage. A rival airman, not wishing to be outdone in skill or courage, immediately tries either to repeat the performance or to perform an even more difficult evolution. The result may very well end in disaster, and

FAMOUS AIRMAN KILLED

is seen on most of the newspaper bills.

The daring of some of our professional airmen is notorious. There is one particular pilot, whose name is frequently before us, whom I have in mind when writing this chapter. On several occasions I have seen him flying over densely–packed crowds, at a height of about two hundred feet or so. With out the slightest warning he would make a very sharp and almost vertical dive. The spectators, thinking that something very serious had happened, would scatter in all directions, only to see the pilot right his machine and jokingly wave his hand to them. One trembles to think what would have been the result if the machine had crashed to earth, as it might very easily have done. It is interesting to relate that the risks taken by this pilot, both with regard to the spectators and himself, formed the subject of comment, and, for the future, flying over the spectators' heads has been strictly forbidden.

From 1909 to 1913 about 130 airmen lost their lives in Germany, France, America, and the British Isles, and of this number the British loss was between thirty and forty. Strange to say, nearly all the German fatalities have taken place in air–ships, which were for some years considered much safer than the heavier–than–air machine.

CHAPTER XLVIII. Some Technical Terms used by Aviators

Though this book cannot pretend to go deeply into the technical side of aviation, there are certain terms and expressions in everyday use by aviators that it is well to know and understand.

First, as to the machines themselves. You are now able to distinguish a monoplane from a biplane, and you have been told the difference between a TRACTOR biplane and a PROPELLER biplane. In the former type the screw is in front of the pilot; in the latter it is to the rear of the pilot's seat.

Reference has been previously made to the FUSELAGE, SKIDS, AILERONS, WARPING CONTROLS, ELEVATING PLANES, and RUDDER of the various forms of air–craft. We have also spoken of the GLIDING ANGLE of a machine. Frequently a pilot makes his machine dive at a much steeper gradient than is given by its natural gliding angle. When the fall is about one in six the glide is known as a VOL PLANE; if the descent is made almost vertically it is called a VOL PIQUE.

In some cases a PANCAKE descent is made. This is caused by such a decrease of speed that the aeroplane, though still moving forward, begins to drop downwards. When the pilot finds that this is taking place, he points the nose of his machine at a much steeper angle, and so reaches his normal flying speed, and is able to effect a safe landing. If he were too near the earth he would not be able to make this sharp dive, and the probability is that the aeroplane would come down flat, with the possibility of a damaged chassis. It is considered faulty piloting to make a pancake descent where there is ample landing space; in certain restricted areas, however, it is quite necessary to land in this way.

A far more dangerous occurrence is the SIDE–SLIP. Watch a pilot vol–planing to earth from a great height with his engine shut off. The propeller rotates in an irregular manner, sometimes stopping altogether. When this happens, the skilful pilot forces the nose of his machine down, and so regains his normal flying speed; but if he allowed the propeller to stop and at the same time his forward speed through the air to be considerably diminished, his machine would probably slip sideways through the air and crash to earth. In many cases side–slips have taken place at aerodromes when the pilot has been rounding a pylon with the nose of his machine pointing upwards.

When a machine flies round a corner very quickly the pilot tilts it to one side. Such action as this is known as BANKING. This operation can be witnessed at any aerodrome when speed handicaps are taking place.

Since upside–down flying came into vogue we have heard a great deal about NOSE DIVING. This is a headlong dive towards earth with the nose of the machine pointing vertically downwards. As a rule the pilot makes a sharp nose dive before he loops the loop.

Sometimes an aeroplane enters a tract of air where there seems to be no supporting power for the planes; in short, there appears to be, as it were, a HOLE in the air. Scientifically there is no such thing as a hole in the air, but airmen are more concerned with practice than with theory, and they have, for their own purposes, designated this curious phenomenon an AIR POCKET. In the early days of aviation, when machines were far less stable and pilots more quickly lost control of their craft, the air pocket was greatly dreaded, but nowadays little notice is taken of it.

A violent disturbance in the air is known as a REMOUS. This is somewhat similar to an eddy in a stream, and it has the effect of making the machine fly very unsteadily. Remous are probably caused by electrical disturbances of the atmosphere, which cause the air streams to meet and mingle, breaking up into filaments or banding rills of air. The wind—that is, air in motion—far from being of approximate uniformity, is, under most ordinary conditions, irregular almost beyond conception, and it is with such great irregularities in the force of the air streams that airmen have constantly to contend.

CHAPTER XLIX. The Future in the Air

Three years before the outbreak of the Great War, the Master–General of Ordnance, who was in charge of Aeronautics at the War Office, declared: "We are not yet convinced that either aeroplanes or air–ships will be of any utility in war".

After four years of war, with its ceaseless struggle between the Allies and the Central Powers for supremacy in the air, such a statement makes us rub our eyes as though we had been dreaming.

Seven years—and in its passage the air encircling the globe has become one gigantic battle area, the British Isles have lost the age–long security which the seas gave them, and to regain the old proud unassailable position must build a gigantic aerial fleet— as greatly superior to that of their neighbours as was, and is, the British Navy.

Seven years—and the monoplane is on the scrap–heap; the Zeppelin has come as a giant destroyer—and gone, flying rather ridiculously before the onslaughts of its tiny foes. In a recent article the editor of The Aeroplane referred to the erstwhile terror of the air as follows: "The best of air–ships is at the mercy of a second–rate aeroplane". Enough to make Count Zeppelin turn in his grave!

To–day in aerial warfare the air–ship is relegated to the task of observer. As the "Blimp", the kite–balloon, the coast patrol, it scouts and takes copious notes; but it leaves the fighting to a tiny, heavier–than–air machine armed with a Lewis gun, and destructive attacks to those big bomb–droppers, the British Handley Page, the German Gotha, the Italian Morane tri–plane.

The war in the air has been fought with varying fortunes. But, looking back upon four years of war, we may say that, in spite of a slow start, we have managed to catch up our adversaries, and of late we have certainly dealt as hard knocks as we have received. A great spurt of aerial activity marked the opening of the year 1918. From all quarters of the globe came reports, moderate and almost bald in style, but between the lines of which the average man could read word–pictures of the skill, prowess, and ceaseless bravery of the men of the Royal Flying Corps and Royal Naval Air Service. Recently there have appeared two official publications [1], profusely illustrated with photographs, which give an excellent idea of the work and training of members of the two corps. Forewords have been contributed respectively by Lord Hugh Cecil and Sir Eric Geddes, First Lord of the Admiralty. These publications lift a curtain upon not only the activities of the two Corps, but the tremendous organization now demanded by war in the air.

[1] The Work and Training of the Royal Flying Corps and The Work and Training of the Royal Naval Air Service.

All this to–day. To–morrow the Handley Page and Gotha may be occupying their respective niches in the museum of aerial antiquities, and we may be all agog over the aerial passenger service to the United States of America.

For truly, in the science of aviation a day is a generation, and three months an eon. When the coming of peace turns men's thoughts to the development of aeroplanes for commerce and pleasure voyages, no one can foretell what the future may bring forth.

At the time of writing, air attacks are still being directed upon London. But the enemy find it more and more difficult to penetrate the barrage. Sometimes a solitary machine gets through. Frequently the whole squadron of raiding aeroplanes is turned back at the coast.

As for the military advantage the Germans have derived, after nearly four years of attacks by air, it may be set down as practically nil. In raid after raid they missed their so–called objectives and succeeded only in killing noncombatants. Far different were the aim and scope of the British air offensives into Germany and into country occupied by German troops. Railway junctions, ammunition dumps, enemy billets, submarine bases, aerodromes—these were the targets for our airmen, who scored hits by the simple but dangerous plan of flying so low that misses were almost out of the question.

"Make sure of your objective, even if you have to sit upon it." Thus is summed up, in popular parlance, the policy of the Royal Flying Corps and Royal Naval Air Service. And if justification were heeded of this strict limitation of aim, it will be found in the substantial military losses inflicted upon the enemy results which would never have been attained had our airmen dissipated their energies on non–military objectives for the purpose of inspiring terror in the civil population.

LaVergne, TN USA
08 September 2009
157225LV00006B/63/A

9 781419 172069